Rock Mechanics Principles in Engineering Practice

CIRIA, the Construction Industry Research and Information Association, is an independent non-profit-distributing body which initiates and manages research and information projects on behalf of its members. CIRIA projects relate to all aspects of design, construction, management, and performance of buildings and civil engineering works. Details of other CIRIA publications, and membership subscriptions rates, are available from CIRIA at the address below.

This CIRIA Ground Engineering Report was written by Dr J. A. Hudson under contract to CIRIA and with the guidance of the project steering group:

S. J. Barnes BSc CEng MICE FGS   Department of Transport
D. Brook BSc PhD FGS   Department of the Environment
A. W. Davies BSc   UK Nirex Ltd
J. N. Davies BSc MSc DIC CEng MICE   Tarmac Construction Ltd
T. H. Douglas FICE FIStructE FIWEM FGS   James Williamson and Partners
    MConsE (Chairman)
G. D. Matheson BSc PhD CEng MIMM FGS TRRL – Scottish branch
S. R. Newson BSc FIMinE   British Coal
P. Tedd BSc PhD   Building Research Establishment
M. F. Warner BSc MSc DIC CEng MICE   Property Services Agency

The project was funded by:

Department of Transport   North of Scotland Hydro-Electric Board
    TRRL – Scottish branch   Property Services Agency
James Williamson and Partners   UK Nirex Ltd
Mott, Hay and Anderson

The photograph on the front cover shows discontinuities in Mountsorrel granodiorite, Buddon Wood quarry, UK

CIRIA's Research Manager for Ground Engineering is F. M. Jardine.

CIRIA
6 Storey's Gate
London SW1P 3AU
Tel. 01-222 8891
Fax. 01-222 1708

CIRIA Ground Engineering Report: Underground Construction

Rock Mechanics Principles in Engineering Practice

J. A. Hudson PhD DSc MICE FIMM CEng
Imperial College of Science, Technology and Medicine, University of London

Construction
Industry
Research and
Information
Association

Butterworths
London   Boston   Singapore
Sydney   Toronto   Wellington

PART OF REED INTERNATIONAL P.L.C.

First published 1989

© **CIRIA, 1989**

British Library Cataloguing in Publication Data

Hudson, J.A. (John Anthony), *1940–*
    Rock mechanics principles in engineering practice.
    1. Rocks. Mechanics
    I. Title   II. Construction Industry Research
    and Information Association
    III. Series 624.1'5132

ISBN 0-408-03081-X

Library of Congress Cataloging-in-Publication Data

Hudson, J.A. (John A.)
    Rock mechanics principles in engineering practice / J.A. Hudson.
        p.   cm. – – (CIRIA ground engineering report)
    Bibliography: p.
    Includes index.
    ISBN 040803081X:
    1. Rock mechanics.   I. Title.   II. Series.
TA706.H83   1989
624.1'5132—dc19   88-31790

Typeset by Saxon Printing Ltd., Derby
Printed and bound by Hartnoll Ltd., Bodmin, Cornwall

8.2.90

Contents

Acknowledgements   **1**

1   Introduction   **2**

1.1   Rock engineering projects   4
1.2   The geological background   6
1.3   The nature of intact rock and rock masses   8

2   Rock mechanics principles   **10**

2.1   Deformability, strength and failure of intact rock   12
2.2   Discontinuities and stereographic projection   15
2.3   Natural and induced stress   18
2.4   Deformability, strength and failure of rock masses   20
2.5   Permeability of rock   22
2.6   Anisotropy   24
2.7   Inhomogeneity   25
2.8   Representative elemental volume   26
2.9   Continuum and discontinuum methods of analysis   27
2.10   Interactions in rock engineering   28

3   Measurement of rock behaviour and rock properties   **29**

3.1   Index tests   29
3.2   Measurement of discontinuity characteristics   31
3.3   Measurement of displacement   34
3.4   Measurement of stress   36
3.5   Rock mass classification schemes   38

4   Excavation and support   **40**

4.1   Blasting   41
4.2   Pre-splitting   42
4.3   Mechanised excavation   43
4.4   Block theory   44
4.5   The ground response curve   45
4.6   Rock reinforcement   46
4.7   Rock support   47

5   Established applications   **48**

5.1   Foundations   49
5.2   Surface excavations and slope stability   51
5.3   Underground excavations   54
5.4   Application example: tunnel portal design   62

6   New applications   **64**

6.1   Geothermal energy   64
6.2   Radioactive waste disposal   66
6.3   The use of underground space   68

7   Concluding remarks   **70**

Index   **71**

Acknowledgements

The following individuals and bodies provided photographs or other illustrative material.

Babtie, Shaw and Morton
Building Research Establishment
Calor Gas Ltd
Conoco Limited
Fairclough Civil Engineering Ltd
Dr R. J. Fowell
James Williamson and Partners
Dr J. C. S. Long
Dr G. D. Matheson
Mr L. A. Merino
Dr Y. Ohnishi
Dr R. J. Pine
Redland Aggregates Ltd
Ms M. Richardson
Scottish Development Department
TRRL
TRRL – Scottish Branch
Tarmac Construction Ltd
Dr P. Tedd
UK Nirex Ltd

Special thanks are also due to: Dr R. W. Poole, who initiated the research project leading to this book; Mr F. M. Jardine, who saw the project through to completion; Ms J. M. Orebi Gann, who prepared it for publication; and Mr A. D. Boddy, who supplied the line drawings.

1
Introduction

Throughout history, underground openings have been used as shelter for the living and the dead. Man has searched avidly for minerals, and has tunnelled for storage, for water, to make roads, and to carry or escape siege. All types of civil engineering project can involve rock excavation, some at the surface, some at depth. Underground space is now being considered for new purposes, such as radioactive waste disposal and the storage of superconductive magnetic energy. Although rock engineering is based on thousands of years of experience, it is only in the last few decades that it has been supported by scientific theory and measurement.

Mechanics is the study of the response of a material to an applied disturbance. In rock engineering, this involves understanding the material properties and the pre-existing boundary conditions (such as *in-situ* stress) as well as the nature of the disturbance itself. Rock mechanics, with its associated principles and applications, has been a discipline in its own right for about 25 years – 1987 was the silver jubilee of the International Society for Rock Mechanics, while 1988 was that of the International Journal of Rock Mechanics and Mining Sciences.

Now that the subject has attained this level of maturity, there is increasing motivation to apply the principles of rock mechanics in engineering practice. There are about 20 major text books devoted to a variety of aspects of rock mechanics and rock engineering, but there is as yet no guide which sets rock mechanics principles in the direct context of engineering practice.

It should be emphasised that this book is an introduction to rock mechanics principles in engineering practice. It is intended not as a design guide but as an overview of the subject of rock mechanics, for engineers who are involved with projects on or in rock masses.

The diagram opposite indicates in context the topics which have to be considered for the different types of surface and underground rock engineering, and how solutions are approached. The inner ring represents the study of individual subjects (discussed mainly in Sections 1 to 4). In rock, different factors may interact with each other; the middle ring represents this stage of study, which is also illustrated by the matrix diagram in Section 2.10 and the construction example in Section 5.4. The outer ring represents the total rock engineering project, i.e. the overall resolution of the problem.

Two of the main topics are excavation and support; the photographs below illustrate the process of excavation.

Blasting in Chile

Tunnel boring machine

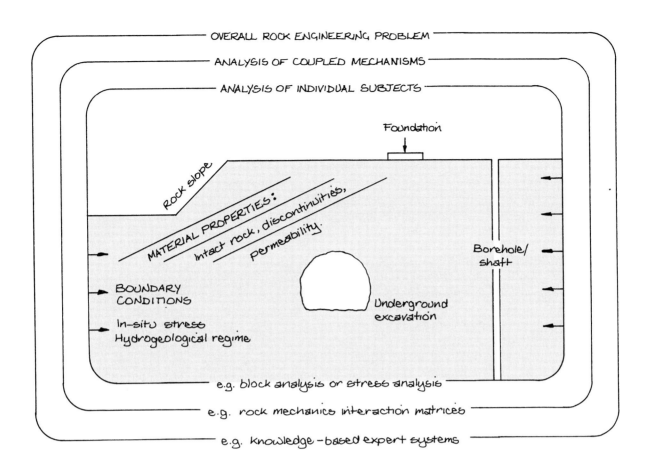

OVERALL ROCK ENGINEERING PROBLEM

ANALYSIS OF COUPLED MECHANISMS

ANALYSIS OF INDIVIDUAL SUBJECTS

Foundation

Rock slope

MATERIAL PROPERTIES: Intact rock, discontinuities, permeability.

Borehole/ shaft

BOUNDARY CONDITIONS

In-situ stress Hydrogeological regime

Underground excavation

e.g. block analysis or stress analysis

e.g. rock mechanics interaction matrices

e.g. knowledge-based expert systems

Each rock engineering project is different, but the principles of rock mechanics are relevant to all the design and construction activities which make up the project. This book provides a simple introduction to these principles and their relevance to rock engineering. It is not a manual of rock engineering, which is a partly empirical discipline requiring practice, observation and correction for success. Empirical skills improve with understanding, however, and the purpose of this book is to aid understanding.

Both the established study of rock engineering and the relatively new science of rock mechanics are developing rapidly as each interacts with the other, and as man makes more demands on the earth's resources and available space. These new challenges will in turn stimulate further development in rock technology.

The photographs below show the need to consider support to ensure the stability of the finished excavation.

Construction of a water tunnel in the UK

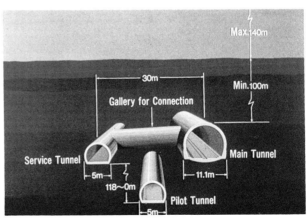

Geometry of the Seikan tunnel, Japan

1.1 Rock engineering projects

Seven categories of rock engineering activity are shown on this and the opposite page. They include mining, quarrying and civil engineering applications, as well as the creation of useable space below ground. Some categories are relevant to more than one type of activity; rock slopes, for example, may be formed for a highway or during open-cast mining. Tunnels and shafts have the same purpose of access whether they are for a deep mine or for a mass transit system, but the design philosophies can be quite different. Similarly, mining and creating a permanent cavern both involve excavation and support, but they have entirely different objectives: in mining only what has economic value is excavated, and collapse is prevented only to maintain access to the working face; whereas with a permanent excavation the cavern must not collapse, its shape or volume is the purpose of excavation, and the excavated material is run to spoil.

The first three categories – foundations, rock slopes, shafts and tunnels – are established disciplines. For caverns, however (although many have been constructed) new civil engineering applications are being developed, such as underground storage. The different methods of mining are areas to which the theory of rock mechanics has been little applied, and there is also little precedent practice (i.e. practice based on established methods) relating to the newer subjects of geothermal energy and radioactive waste disposal. In all these applications, whether established or new, the principles of rock mechanics are directly relevant.

1.1.1 Foundations

Rock is usually an excellent foundation material, but near-surface rock can be significantly fractured: it is always necessary to establish the competence of the rock to bear the required load at acceptable levels of deformation or settlement.

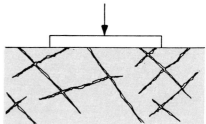

Key references

1. GOODMAN, R. E. (1980) Introduction to rock mechanics. John Wiley and Sons (New York)
2. GOODMAN, R. E. and GEN, H. S. (1985) Block theory and its application to rock engineering. Prentice-Hall (New Jersey)

1.1.2 Rock slopes

There are four basic mechanisms for rock slope failure: plane, wedge, direct toppling, and flexural toppling. The potential for failure in any of these modes can be easily identified using rock mechanics methods. The need and scope for a more detailed analysis can then be assessed.

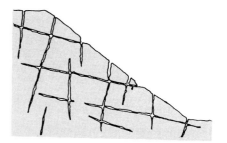

Key references

1. HOEK, E. and BRAY, J. W. (1981) Rock slope engineering. Institute of Mining and Metallurgy, London
2. MATHESON, G. D. (1983) Rock stability assessment in preliminary site investigations – graphical methods. TRRL LR 1039, Transport and Road Research Laboratory, Crowthorne, Berks

1.1.3 Shafts and tunnels

The stability of shafts and tunnels depends on rock structure, rock stress, groundwater flow, and construction technique. The fundamental principles of mechanics provide extremely useful guidelines for stability assessment.

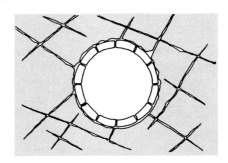

Key references

1. MEGAW, T. M. and BARTLETT, J. V. (1981, vol 1 and 1982, vol 2) Tunnels: planning, design, construction. Ellis Horwood (Chichester, Sussex)
2. BIENIAWSKI, Z. T. (1984) Rock mechanics design in mining and tunnelling. Balkema (Rotterdam)

1.1.4 Caverns – use of underground space

Rock joints have a major influence on the design and construction of large caverns. Methods to reinforce and support the rock are based on the principles of ground movement resulting from excavation.

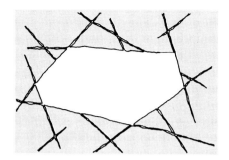

Key references

1. HOEK, E. and BROWN, E.T. (1980) Underground excavations in rock. Institution of Mining and Metallurgy, London
2. SAARI, K. (1986) [ed.] Large rock caverns, Proceedings of Large Rock Caverns Conference, Helsinki

1.1.5 Mining

There is a huge variety of mining geometries, but in all cases the mining methods are designed to extract the mineral with minimum artificial support.

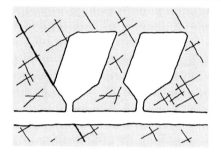

Key references

1. BRADY, B. H. G. and BROWN, E. T. (1985) Rock mechanics for underground mining. George Allen and Unwin (London)
2. PENG, S. S. (1985) Coal mine ground control, 2nd ed., John Wiley and Sons (New York)

1.1.6 Geothermal energy

In extracting geothermal energy, cold water is pumped down one borehole, to pass through fractures in a hot rock reservoir and exit from a second borehole. The optimal configuration for a production system depends on the interactions between rock joints, *in-situ* stress, water flow, temperature and time.

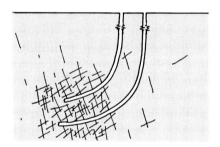

Key reference

1. PINE, R. J., LEDINGHAM, P. and MERRIFIELD, C. M. (1983) In situ stress measurement in the Carnmenellis granite – II. Hydrofracture tests at Rosemanowes Quarry to depths of 2000m. *Int. J. Rock Mech. Min. Sci.* **20**, April 1983, 63-72

1.1.7 Radioactive waste disposal

The aim is to isolate the waste so that unacceptable quantities of radionuclides do not return to the biosphere. Predicting the safety of a repository requires an understanding of all the factors listed above for geothermal energy, as well as others such as radionuclide sorption by rock fracture surfaces.

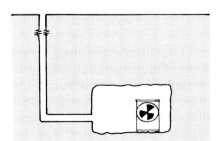

Key references

1. MILNES, A. G. (1985) Geology and radwaste. Academic Press (New York)
2. RADIOACTIVE WASTE (PROFESSIONAL) DIVISION OF THE DEPARTMENT OF THE ENVIRONMENT (1986) Assessment of best practicable environmental options (BPEOs) for management of low- and intermediate-level solid radioactive wastes. HMSO (London)
3. CHAPMAN, N. A. and MCKINLEY, I. G. (1987) The geological disposal of nuclear waste. John Wiley and Sons (London)

1.2   The geological background

Working the anticline in a limestone quarry

Rock is not made to specification; in rock engineering, therefore, where the rock itself is both the construction material and the structure, its properties have to be established from laboratory and field tests. As is clear from the photograph above, the rock mass may have important structural features and will be fractured. It must therefore be recognised as a discontinuous material, which can be expected to have different properties in different locations and directions. What is encountered is not a newly fabricated material, but one which has been subjected to often severe mechanical, thermal and chemical actions over millions of years.

In order to predict how rock will behave as an engineering material, certain sets of properties have to be determined, e.g. those of:

– the intact rock
– the fractures
– the whole rock mass.

The relative importance of these different properties depends upon the particular engineering applications, but it is vitally important that the structural geology is fully appreciated, including stratification, folding, faulting, shearing, jointing, etc, as well as lithology (the rock type).

Within the geological context, and given the properties of the intact rock and the fractures, the other main design considerations are the stress state in the rock, hydrogeological conditions and the construction procedures.

The term **intact rock** refers to rock which has no through-going fractures significantly reducing its tensile strength. Its properties are most important in rock excavation. It is usually characterised by density (unit mass), deformability (Young's modulus and Poisson's ratio), and strength (unconfined compressive strength, cohesion and angle of friction).

It is difficult, however, to provide a cogent description of intact rock where its properties vary widely over the area of engineering interest. Moreover, even simple indicators of a property, e.g. Young's modulus, have to be qualified if the rock has different properties in different directions. In the most extreme case, the rock can have 21 elastic constants rather than the two elastic properties (Young's modulus and Poisson's ratio) of an isotropic material.

This leads to two quite different approaches to specifying intact rock properties:

– direct measuring of fundamental properties such as deformability;
– index testing as a comparative indication of intact rock quality.

The former is more relevant to theoretical analysis; the latter is easier and cheaper to undertake. It is therefore possible to conduct many index tests on relatively large quantities of rock, thus providing a better overall assessment of rock mass quality, rather than performing fundamental tests at isolated locations.

All rock contains **fractures**. These are of varying types and occur on varying scales (from microfissures, through fissures, joints and bedding planes, to faults). The origin and interpretation of the structural geology of these fractures can often be of great assistance to engineers in indicating the mechanical structure and properties. For example, many sedimentary rocks are divided by major discontinuities into large blocks. The discontinuities are often uniform in orientation and persistence, but the joints which fracture these blocks internally are frequently much less consistent in direction and extent.

The term **discontinuity** is used in rock engineering for all such types of fracture to indicate that the rock is not continuous, unlike the intact rock described above which is mechanically continuous. Clearly the nature, location and orientation of discontinuities profoundly affect most of the rock properties (deformability, strength, permeability, etc.), and therefore the rock engineering applications. The photograph below shows a rock mass containing discontinuities which will clearly dominate its mechanical and hydraulic properties.

Discontinuities in the Mountsorrel granodiorite, Buddon Wood quarry, UK

Key references

1. HOBBS, B. E., MEANS, W. D. and WILLIAMS, P. F. (1976) An outline of structural geology. John Wiley and Sons (New York)
2. DAVIS, G. H. (1984) Structural geology of rocks and regions. John Wiley and Sons (New York)
3. BLYTH, F. G. H. and DE FREITAS, M. H. (1984) A geology for engineers, 7th ed. Edward Arnold (London)

1.3 The nature of intact rock and rock masses

The essence of a rock mechanics problem is shown in the diagram below. A rock mass (consisting of intact rock and discontinuities), which could well be under significant stress, is being disturbed by engineering activity. This points to another major difference between rock engineering and many other types of engineering: it is the removal of material that is the engineering activity. In other words, the stresses in the rock are being redistributed rather than directly applied.

The design philosophy will be governed by the objective – whether this is the removal of material to create a permanent structure, as in civil engineering, or to obtain the material itself, as in mining engineering.

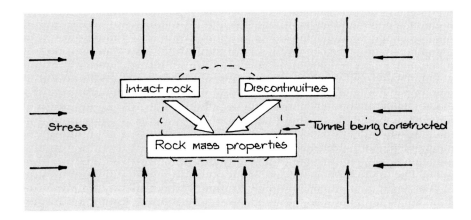

1.3.1 Intact rock

The history of testing intact rock has been dictated to a large extent by sampling methods. Rock obtained remotely is in the form of diamond-drilled cores, which can easily be formed into regular cylinders and disc samples for testing. Probably the most widely used rock property is the unconfined compressive strength, σ_c. This is relatively easy to measure directly (see Section 2.1) and can also be estimated from the most widely used index test, the point load test (see Section 3.1).

The diagram on the right illustrates ranges of rock strength. As a rule of thumb, 1 MPa can be considered as the interface between soil and rock strengths. The full range is extremely wide and so the variation in strengths over a site can be important, e.g. as a factor in choosing excavation techniques, where a mean strength would be misleading.

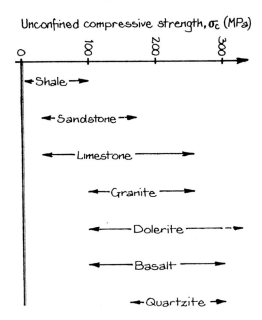

Key references

1. JAEGER, J. C. and COOK, N. G. W. (1979) Fundamentals of rock mechanics, 3rd ed. Chapman and Hall (London)
2. VUTUKURI, V. S., LAMA, R. D. and SALUTA, S. S (1974) Handbook on mechanical properties of rocks, Vol 1, Trans Tech Publications

1.3.2 Discontinuities

Discontinuities influence almost all the engineering properties and behaviour of rock, but their configuration cannot as yet be predicted nor their character adequately described. Analytical techniques for modelling the effects of discontinuities are still in their infancy; it is usually necessary to treat the discontinuity pattern as though each discontinuity is planar and through-going, and as consisting of regular sets of parallel discontinuities.

To assess the effects of discontinuities on the deformability of rock, it is usual to assume that they have linear deformation characteristics in compression and shear (by definition they cannot sustain tension). When assessing the permeability of a fractured rock, the first step might be to assume a constant aperture (width across the discontinuity) for all the discontinuities, but a more detailed and refined representation of the discontinuities would be needed for a more realistic analysis.

Special presentational and statistical methods are used to describe and represent discontinuity configurations. These methods are particularly relevant to engineering problems such as the stability of a rock face, and they are described in Sections 2.2 to 2.9 inclusive.

Key references

1. BROWN, E. T. [ed.] (1981) Rock characterization, testing and monitoring: ISRM suggested methods. Pergamon Press (Oxford)
2. HUDSON, J. A. and PRIEST, S. D. Rock discontinuities in engineering (in preparation)

Discontinuity sets

1.3.3 Rock masses

Intact rock properties and discontinuity configurations are important in excavation, but in different ways, depending on the purpose and method of excavation. Examples of these differences are:

- Rock cutting with a tunnel-boring machine (intact rock)
- Rock ripping of overburden (discontinuities)
- Quarrying dimension stone for buildings (discontinuities and intact rock).

For support, the local discontinuities are important. When the whole rock mass has to be considered, the combined effect of the intact rock and the discontinuities, i.e. the properties of the discontinuum, is taken into account. Although almost any geometrical configuration of discontinuities is possible, in sedimentary rocks there are often three mutually perpendicular sets of discontinuities: the bedding planes, and two sets of joints, as illustrated above. In this case, the procedures for sampling, measuring and modelling the rock mass are reasonably straightforward. Other cases are more complex, however, especially when the properties and behaviour are governed by the relation between the scale of the project and the average spacing between the discontinuities (see Section 2.4.1). But even in these cases, considerable guidance can be obtained from the techniques described in Sections 2.2 to 2.9.

Key references

1. FARMER, I. W. (1983) Engineering behaviour of rocks, 2nd ed. Chapman and Hall (London)
2. BIENIAWSKI, Z. T. (1984) Rock mechanics design in mining and tunnelling. Balkema (Rotterdam)

2

Rock mechanics principles

The most important principles of rock mechanics are explained in this Section. They fall into nine main categories, and lead to a tenth dealing with the interactions between the principles in the context of rock engineering.

Before discussing the ten categories, it is useful to examine the differences between the disciplines of rock mechanics and soil mechanics, and to explain what is meant by the expression 'interactions in rock engineering'.

Soil mechanics and rock mechanics
Excavation, whether cutting a slope, tunnelling, making a storage tank or forming a foundation, can involve either soil or rock or both. The engineering requirements of stability and competence have to be achieved for both materials, but the theory relating to the behaviour of each differs in emphasis.

The transition from rock to soil mechanics can be visualised by progressively fracturing a rock mass, bearing in mind that rock is fractured *in situ*, whereas individual soil grains have usually been transported. Intact rock, although composed of discrete grains or individual crystals, is so interlocked or cemented as to be a coherent material, but as it is progressively fractured it becomes increasingly granular. An arbitrary but convenient division between rock and soil is a compressive strength of 1 MPa.

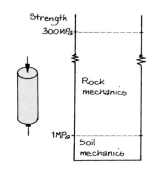

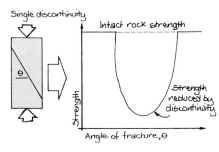

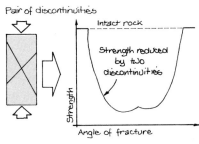

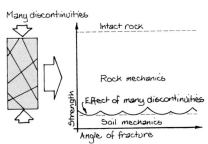

This transition from intact rock to highly fractured rock is illustrated by the way compressive strength decreases as the fracturing in a sample increases. The diagrams demonstrate the influence of discontinuities on strength: firstly of a single fracture at different inclinations to the axis of the sample and, secondly, of two fractures. Where there are many such discontinuities, the rock is weakened in all directions, as indicated by the cusped curve in the third diagram.

The two most important differences between rock mechanics and soil mechanics are that in a rock:

- The discontinuities can be on the same scale as the excavation (see Section 2.4.1); this is not the case for soils because the grains are several orders of magnitude smaller than the engineering dimensions.
- The discontinuities are not necessarily fully interconnected, i.e. a discontinuity does not necessarily form the outer boundary of a rock block, nor is it always a channel for the free flow of water.

The discontinuities are the factor which causes rock mechanics to be a discipline in its own right (even though much of the stress analysis side of ground engineering applies equally to rock and soil mechanics). The discontinuities introduce a variety of additional considerations which are very significant. The term rock structure is used to describe the three-dimensional arrangement of discontinuities within a rock mass.

Interactions in rock engineering

All forms of ground engineering are concerned with the interaction of the construction process with the ground. Additionally, there is interaction between the rock properties themselves during all engineering activities. It is instructive at this stage to consider interactions between the three primary characteristics of a rock mass:

- rock structure
- *in-situ* stress
- water flow in the rock mass.

The interactions are illustrated below, using a matrix numbering system, i.e. 1, 2; 2,1, etc., to denote the six possible combinations. Note that interaction 1, 2 is not the same as 2,1, because the effect of rock structure (1) on *in-situ* stress (2) is not the same as that of stress (2) on rock structure (1). Interactions are discussed further in Section 2.10, but there are two points worth noting at this stage. Firstly, although each of the basic principles of rock mechanics is explained separately in the sections that follow, they should eventually be considered in combination. Secondly, the interaction process is an excellent starting point for studying either a rock engineering project as a whole, or individual elements of the project.

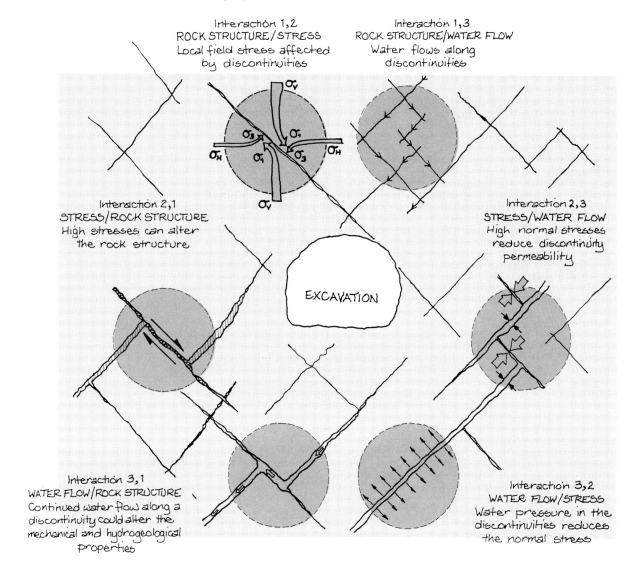

Interaction 1,2
ROCK STRUCTURE/STRESS
Local field stress affected by discontinuities

Interaction 1,3
ROCK STRUCTURE/WATER FLOW
Water flows along discontinuities

Interaction 2,1
STRESS/ROCK STRUCTURE
High stresses can alter the rock structure

Interaction 2,3
STRESS/WATER FLOW
High normal stresses reduce discontinuity permeability

EXCAVATION

Interaction 3,1
WATER FLOW/ROCK STRUCTURE
Continued water flow along a discontinuity could alter the mechanical and hydrogeological properties

Interaction 3,2
WATER FLOW/STRESS
Water pressure in the discontinuities reduces the normal stress

A further complication arises for the range of soft rocks, such as chalk or mudstones, because their properties can change with time after being disturbed by engineering. For example, on exposure to air or water or when subjected to stress changes, the rocks weaken, possibly to below the assumed design strength.

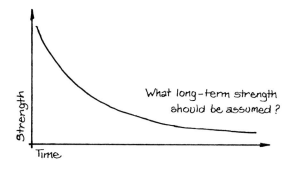

What long-term strength should be assumed?

2.1 Deformability, strength and failure of intact rock

To a large extent, the history of rock testing to determine rock properties has developed from the use of cylindrical specimens – because it is only by diamond core drilling that a coherent, remote rock sample can be obtained. In particular the stiffness, strength and failure behaviour of rock have been studied using regular cylindrical specimens. As discussed below, the most widely used index test (the point load test) has the advantage that it can be used not only on core, either along the axis or across a diameter, but also on irregular samples.

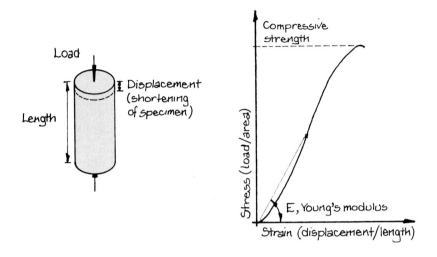

Because the stress/strain plot is usually curved, a variety of methods are used to estimate E, the assumed linear relation between stress and strain.

Until 1966, load/displacement measuring terminated just after the peak strength had been reached, because the rock specimens failed 'explosively'. This explosive failure was thought to be an inherent characteristic of the rock, although it is incompatible with the generally 'peaceful' nature of underground rock failures (the exceptions being rock bursts and coal bumps).

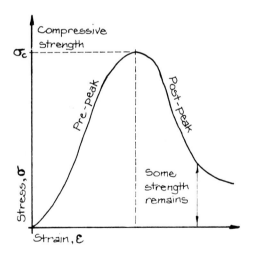

The complete stress/strain curve shown in the diagram represents the total mechanical behaviour in uniaxial compression, from initial loading to complete destruction of the specimen; i.e. the rock is capable of sustaining load beyond the peak, losing strength gradually with increasing strain.

To cause failure, the rock must be taken to the post-peak region; to avoid failure, the rock must remain in the pre-peak region. When the shape of the post-peak region is understood, it is possible to use the post-peak strength in engineering design. This is particularly relevant for the design of mining operations with failing pillars.

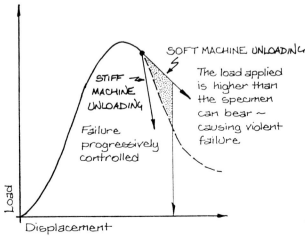

Special testing techniques have to be used to obtain the complete stress/strain curve. In 1966, it was recognised that the stiffness of the testing machine (relative to the slope of the post-peak load/displacement curve) determines whether failure of the specimen is stable or unstable. As shown in the diagram, a 'soft' machine causes sudden failure by the violent release of stored strain energy, i.e. by the testing system itself.

This problem does not generally arise in soil mechanics testing because of the relatively low stiffness of the soil compared with the testing machine.

Machines for testing materials were first developed in the Renaissance period by Leonardo da Vinci (1452-1519), becoming quite sophisticated by late Victorian times as shown in the upper photograph.

For rock mechanics, conventional machines were specially stiffened in order to obtain the complete curves. Stiffness is the force (P) needed to produce a unit change in length (dh)

$$\text{i.e. stiffness} = \frac{\text{force}}{\text{displacement}} = \frac{P}{dh} = \frac{\text{stress} \times \text{area}}{\text{strain} \times \text{length}} = \frac{EA}{h}$$

A testing machine can be stiffened by:

- increasing the modulus of the materials used (E).
- increasing the platen width or area (A)
- reducing the height (h) of the components of the load column.

Victorian testing machine

Servo-controlled testing machine

Although successful, machines of this type are inherently cumbersome and functionally inflexible, so other methods are now used to obtain the complete stress/strain curve. These testing machines are servo-controlled, operating by closed-loop hydraulic control of a particular variable. For example, the complete stress/strain curve can be obtained by programming a linear increase in displacement with time, and utilising the feedback from an axial displacement transducer. In this case the rate of change of stress with time is not constant, but is dictated by the linear increase in applied displacement.

The reason why servo-controlled machines can be used to obtain the full stress/strain curve is that the response time of the system (typically of the order of 1/5000 s) is less than the time required for any significant unstable crack propagation.

Because any variable or, with a computer, any combination of variables can be used as feedback in the closed loop, almost any form of control is possible. In particular, the most sensitive indicators of failure can be used – the lateral strain, for example, in the case of an axially loaded specimen.

An example of a sophisticated servo-controlled testing machine, as used at Imperial College, University of London, is illustrated on the left.

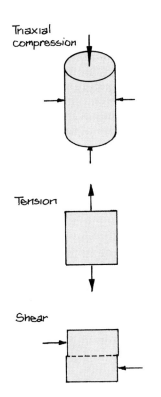

Triaxial compression

Tension

Shear

Just as the complete load-displacement curve can be obtained in uniaxial compression, the equivalent curve can be obtained for any type of loading (e.g. triaxial compression, tension, shear, etc.) using a servo-controlled closed-loop system.

The Young's modulus of rock is measured from the pre-failure stress/strain curve; depending on the type of rock, values vary from less than 1 to more than 100 GPa. For the same type of rock, the modulus values are often reasonably similar, even though strengths can be highly variable. The reason is that at low strains (over which modulus is measured), the whole length of the specimen is being deformed but, near the peak, the strength is controlled by the highly variable flaws in the test specimens.

It is important to remember the difference between measured rock properties and genuine material properties. A true material property does not depend on the specimen geometry or the method of testing, whereas most measured rock properties do. It is because the properties are required for engineering purposes that a degree of standardisation of testing procedures is important. As a corollary, the properties of the intact rock should not be measured until it has been decided exactly why the information is required and which are the best tests to conduct.

It should also be remembered that, although we understand a great deal about intact rock, we still do not understand exactly why intact rock fails. Of the several criteria which are successfully used to define failure, most are empirical, i.e. are based on previous experience, and it is therefore potentially dangerous to apply them in new circumstances.

When the compressive strength of a rock is given without qualification, it is the unconfined, uniaxial-compression, peak-stress strength that is meant. There is no unique value of 'strength' as such because a rock's resistance to load is a function of the loading conditions and state of stress. As mentioned above, the compressive strength of different rocks varies greatly, from about 1 MPa to about 300 MPa, as shown in the diagram below for some British rocks. As yet there is no generally accepted measure of the brittleness of rock, i.e. how to represent the gradual reduction in available strength beyond the peak value.

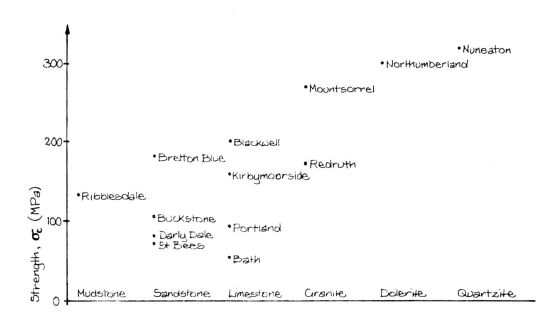

Key references

1. HUDSON, J. A., CROUCH, S. L. and FAIRHURST, C. (1966) Soft, stiff and servo-controlled testing machines; a review with reference to rock failure. *Engineering Geology*, 1966, **6** (3), 155-189
2. BROWN, E. T. [ed.] (1981) Rock characterization, testing and monitoring: ISRM suggested methods. Pergamon Press (Oxford)

2.2 Discontinuities and stereographic projection

2.2.1 Discontinuities

Discontinuities (any type of fracture) are the single most important characteristic of a rock mass because they dominate the modulus, strength and permeability values. They often have a significant effect on the transmission and relief of stress (and hence the stress distribution) in the rock, and thus on the excavation and support methods required.

Discontinuities vary in their geometrical and mechanical characteristics. For engineering, it is necessary to understand and evaluate these, using standard methods of description. Ten characteristics of discontinuities are listed in the ISRM Suggested Methods (see Key reference 1): orientation, spacing, persistence, roughness, wall strength, aperture, filling, seepage, number of sets, and block size. Each of these can have a significant effect on the behaviour of the discontinuities and on the rock mass properties. These characteristics are discussed in greater detail in Section 3.2.

An open discontinuity causes the rock to have zero tensile strength at that location; and, similarly, the presence of discontinuities significantly reduces both the compressive and shear stiffnesses of a rock mass.

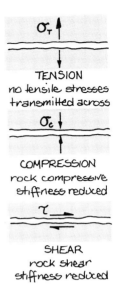

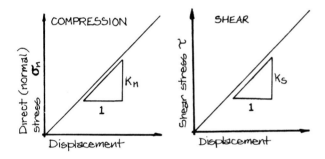

By considering a single discontinuity under compressive or shear loading, the concept of normal stiffness, K_n, and shear stiffness, K_s, can be developed, as illustrated in the two diagrams on the left. Note that the stiffnesses are stress/displacement: the load can be scaled by the area, but the displacement cannot be scaled by the length. In practice, however, these curves are not linear, i.e. the stiffness, K, is not a constant value.

The Young's moduli relevant to a fractured rock mass can be estimated from these stiffnesses and represented by curves such as those in the bottom diagram. The modulus decreases as the discontinuities become more frequent and as the normal stiffness, K_n (for example), is reduced.

All the ten discontinuity attributes affect the modulus of the rock mass to a greater or lesser extent; because of this complexity, it is usual to estimate *in-situ* moduli by some form of index or classification system. Typically, the *in-situ* modulus is about one-tenth of the laboratory-measured value, because of the presence of discontinuities.

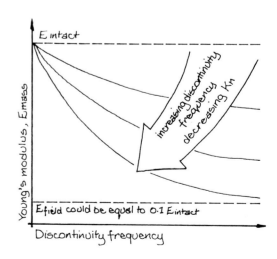

Key reference

1. BROWN, E. T. [ed.] (1980) Rock characterization, testing and monitoring: ISRM suggested methods. Pergamon Press (Oxford)

15

2.2.2 Stereographic projection

Stereographic projection is the two-dimensional representation of three-dimensional discontinuity orientations, i.e. the dips and dip directions of the discontinuity planes. For engineering purposes, lower-hemisphere projection is the most widely used method. The principles are outlined below to show how to read and understand a stereoplot, and to illustrate the value of the technique. Many engineers, because they do not usually take discontinuity measurements and so do not practise stereographic techniques, are unfamiliar with these graphic representations of rock geometry and how they can be used. The key references listed below are strongly recommended; they are models of clarity and provide detailed guidance on the use of stereographic techniques.

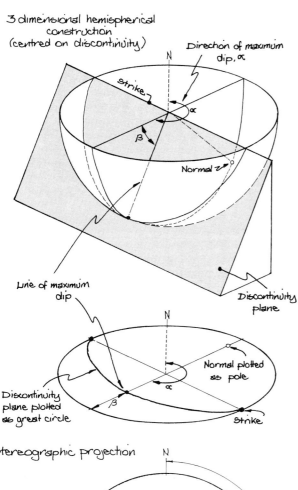

3 dimensional hemispherical construction (centred on discontinuity)

Direction of maximum dip, α

Strike

Normal z

Line of maximum dip

Discontinuity plane

Normal plotted as pole

Discontinuity plane plotted as great circle

Strike

Stereographic projection

Pole

90°

Maximum dip

Strike

dip angle, β

Azimuth (compass bearing) or dip direction, α

A discontinuity plane in space can be conveniently characterised by either of two lines:

– the line of maximum dip in the plane;
– the line perpendicular to the plane, called the normal.

These lines are plotted on the stereographic projection as points. The line of maximum dip in the discontinuity plane is represented by the point at the compass bearing (the direction of the dip) which is at a distance inwards from the perimeter equivalent to the angle of maximum dip. The normal – the line perpendicular to the discontinuity plane – lies on the same diameter of the projection and is the point at a distance equivalent to 90° from the point of maximum dip.

The strike line, which is the line through the discontinuity plane with a dip of zero, is marked by the two points on the perimeter.

The whole discontinuity plane can then be represented either as a curve, the great circle, as shown on the diagram, or as a point, i.e. the pole. The great circle on the projection is the locus of points each of which represents a line on the plane, each line having an apparent dip direction and apparent dip of the plane. Because we may wish to plot hundreds or thousands of discontinuities, it is more convenient to plot the pole points rather than the great circles. When this has been done, the rock structure can be represented by contouring the intensity of pole concentrations on the stereogram.

If these contours lie in reasonably distinct regions of the projection, we can consider the rock as being divided by several sets of discontinuities or suites of subparallel discontinuities. For example, most sedimentary rocks usually have three main sets of fractures: the bedding planes parallel to the stratification and two perpendicular sets. It is often useful to simplify the process by using a single point on the projection to represent one set of discontinuities.

Gaining a working understanding of stereographic projection requires practice but, once achieved, it is very helpful for visualising many aspects of the three-dimensional geometry of a rock mass. Some examples are illustrated and described below.

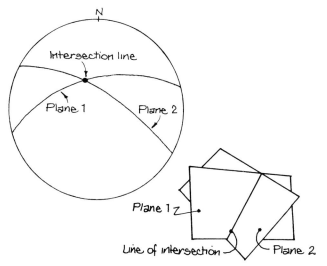

The line of intersection where two discontinuity planes meet is quickly found, as the point of intersection of two great circles on the projection representing the two planes. (Two planes meet at one line in space, two curves meet at one point on the projection.) This is very useful in studying the direction of rock block edges in, for example, wedge failures in rock slopes.

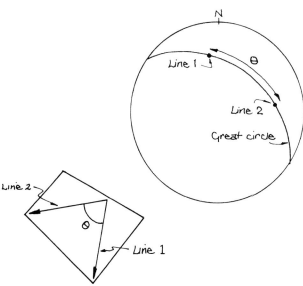

The angle between any two lines in the rock mass can be obtained directly from the projection via the plane or great circle in which both lines lie. This is used, for example, in calculating discontinuity frequency along a borehole in a rock mass, from a knowledge of the set frequencies along the lines perpendicular to the sets.

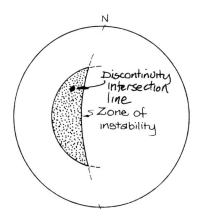

As explained in Section 5.2, criteria for rock slope instability can be represented as shaded zones on the projection. This is because the instability is related to the orientation of the rock slope relative to the orientation of the discontinuities and to the friction angle. By considering the shaded area in relation to the rock structure represented on the stereogram, rapid assessments of the potential for slope instability can be made.

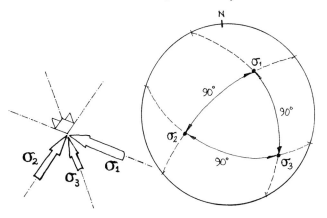

Because any orientational data can be represented on the projection, the method is also useful for plotting stress directions, e.g. those of the three mutually-perpendicular principal stresses which exist at a point in a given rock mass.

Key references

1. PRIEST, S. D. (1985) Hemispherical projection methods in rock mechanics. George Allen and Unwin (London)
2. GOODMAN, R. E. (1980) Introduction to rock mechanics. John Wiley and Sons (New York)
3. MATHESON, G. D. (1983) Rock stability assessment in preliminary site investigations – graphical methods. TRRL LR 1039, Transport and Road Research Laboratory, Crowthorne, Berks

2.3 Natural and induced stress

So far in Section 2 we have discussed the basic material properties. Equally important, however, is the *in-situ* stress field, which is needed to define the boundary conditions for mechanical analysis. In mechanics, we study the effects of applying forces to a material with certain properties. Rock engineering, on the other hand, is concerned mainly with the effect of altering the geometry of a prestressed material, and hence changing the pre-existing stress state when extra loads are applied or when rock is excavated. *In-situ* stress is important for underground engineering, and may also be relevant to some surface engineering.

It is important to recognise that stress is neither a scalar nor a vector quantity, but a tensor quantity.

- Scalar (from the Latin 'to measure'), a quantity with magnitude only, e.g. temperature.
- Vector (from the Latin 'to carry'), a quantity with magnitude and direction, e.g. a force.
- Tensor (from the Latin 'to stretch'), a quantity with magnitude and direction and with reference to the plane it is acting across, e.g. stress or permeability.

There are two main stress components:

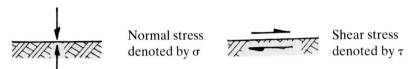

Normal stress denoted by σ Shear stress denoted by τ

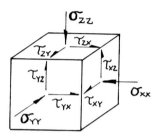

There are nine stress components acting on an element of rock, as illustrated by the cube. These can be listed in matrix form.

Notation: τ_{xy}
First subscript: plane on which τ_{xy} stress acts, given by the direction of its normal (e.g. plane perpendicular to x axis)

Second subscript: direction of stress (e.g. in y direction)

Note: $\tau_{xy} = \tau_{yx}$ etc., otherwise the cube (as shown) would rotate.

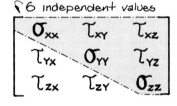

6 independent values

$$\begin{bmatrix} \sigma_{xx} & \tau_{xy} & \tau_{xz} \\ \tau_{yx} & \sigma_{yy} & \tau_{yz} \\ \tau_{zx} & \tau_{zy} & \sigma_{zz} \end{bmatrix}$$

Hence, as indicated by the shaded area, there are six independent components of the stress field – three normal stresses and three shear stresses.

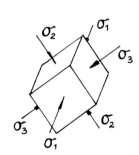

At one specific orientation of the orthogonal planes, the shear stresses on the planes will be zero. The normal stresses on the planes are called the principal stresses; these are illustrated by the cube and listed in the matrix on the right. An example of principal stress directions plotted on a stereonet is shown on the previous page.

An important point regarding rock engineering is that all unsupported excavation surfaces are principal stress planes (because there are no shear stresses acting on them).

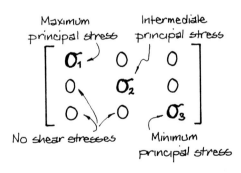

Maximum principal stress   Intermediate principal stress

$$\begin{bmatrix} \sigma_1 & 0 & 0 \\ 0 & \sigma_2 & 0 \\ 0 & 0 & \sigma_3 \end{bmatrix}$$

No shear stresses   Minimum principal stress

Natural vertical stress, p_z
The vertical stress within a rock mass can be estimated from the weight of the overlying material, $p_z \approx 0.027z$ where p_z is in MPa and z is in m.

Rules of thumb to estimate vertical stress are: 1 MPa $\equiv$ 40 m of depth; 1 psi $\equiv$ 1 ft of depth.

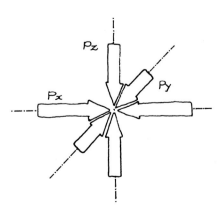

Natural horizontal stresses, p_x and p_y

The horizontal stress within a rock mass is induced by the weight of the overlying material, but the stress field can also be altered by tectonic stresses, erosion and other geological factors.

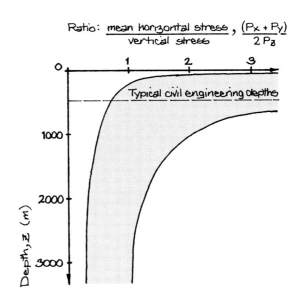

The shaded area in the diagram represents the range of stress fields measured world-wide (see Key reference 1). It shows that the mean of the horizontal stress components is generally greater than the vertical component, at civil engineering depths. The two envelopes represent the upper and lower ratios between the mean horizontal and vertical stresses.

In very soft materials, the principal stresses can approach the condition $\sigma_1 = \sigma_2 = \sigma_3$, which is hydrostatic with no shear stresses. It is the rule rather than the exception that the maximum component of horizontal stress (i.e. whichever of p_x and p_y is the greater) is more than the vertical stress at civil engineering depths in rock.

Induced stress

Construction activities alter the existing stress field; the changed stress is called the induced stress. The diagram shows how the vertical stress component is altered by tunnel construction. No external load has been applied, but the removal of rock results in the surrounding rock having to accommodate a redistributed load.

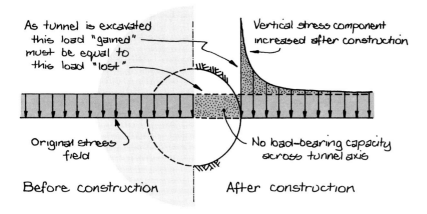

The new stress distributions resulting from excavation are particularly important if they are likely to cause failure. In order to estimate the stress redistribution, e.g. for the design of permanent works underground, it may be necessary to use numerical analyses such as the finite element method, although for simple geometries the estimation methods given in Key reference 1 are often sufficient for engineering purposes.

Effective stress

Any pressure caused by the presence of water in the rock acts equally in all directions and hence this water pressure, u, can be subtracted from all normal components of the stress field, σ, to give the effective stress $\sigma' = \sigma - u$. This concept is fundamental to soil mechanics, but has to be used with caution in rock mechanics because the water tends to flow mainly through rock fractures, which may be sparse and only partially connected.

This subject is currently a very active research area in rock mechanics.

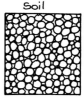

Key references

1. HOEK, E. and BROWN, E.T. (1980) Underground excavations in rock. Institution of Mining and Metallurgy, London
2. STEPHANSSON, O. [ed.] (1986) Rock stress and rock stress measurements, Proceedings of the International Symposium, Centex, Luleå
3. CROUCH, S. L. and STARFIELD, A. M. (1983) Boundary element methods in solid mechanics. George Allen and Unwin (London)

2.4 Deformability, strength and failure of rock masses

A rock mass is basically intact rock containing discontinuities. The properties of the mass depend not only on the properties of the intact material and the discontinuities separately, but also on the way in which they are combined. Thus there are several complications: intact rock is itself inhomogeneous; the discontinuities are diverse in nature (e.g. variations of the ten characteristics in Section 2.2); and their geometrical arrangements are infinitely variable. A further factor relevant to rock engineering is that the rock mass responds to alterations in its state of stress rather than to a superimposed load. Because of these complications, any consideration of the properties of a rock mass has to be based on field tests and to be largely empirical, for example using classification schemes.

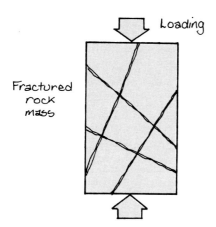

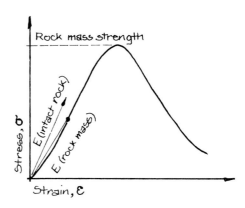

If a rock mass is loaded, its stress/strain curve will not be the same as that of intact rock – its modulus of deformation will be lower than the Young's modulus for the intact rock, and its peak strength will also be lower. But the results will depend on the sample size, because each sample will contain a different discontinuity geometry.

In practice the volume needed for reproducible results is often so large that it would be too costly, if not impracticable, to carry out such a field test. Even measuring deformability may prove impracticable at other than low strains, let alone the full-scale determination of strength or ultimate rock failure. On a small scale, the use of a high-pressure dilatometer or the Goodman borehole loading jack can give useful results, but these will be strongly affected by nearby discontinuities.

Alternatively, or to supplement direct measurements, rock mass classification schemes may be used to assess deformability. The Bieniawski scheme (which is briefly explained in Section 3.5) identifies a rock mass rating (RMR) number based on five attributes of the rock. The maximum posible RMR value (i.e. for the soundest rock) is 100. The *in-situ* modulus of deformability can then be estimated as $E_m \approx 2(RMR) - 100$ (in GPa).

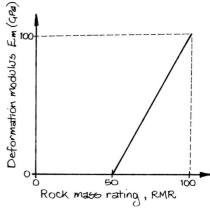

It should be noted that classification systems are based on a specific body of experience; great care should be taken before applying any classification system to new circumstances.

The deformability and strength of rock masses and their engineering implications highlight the differences between rock engineering and other forms of engineering. The material properties are not measured but estimated. The stress is not applied by the engineering but is already present. The rock is neither deformed uniformly nor broken uniformly. As a consequence, two very important factors in rock engineering are the scale of the engineering activity relative to the rock structure, and the depth at which the engineering activity is taking place.

2.4.1 Scale and stability

The effects of scale are important for both surface and underground excavation. Although illustrated here for an underground excavation, the principles apply equally to surface excavation.

In the diagram, the rock mass is considered to have statistically uniform properties. The excavation geometry is much the same; the rock structure is the same; yet the cavern is much less stable than the borehole, because more blocks can fall into it. In other words, stability depends on the ratio of the excavation dimension and the discontinuity spacing. This again emphasises the significance of discontinuities and the reason why rock mechanics and rock engineering are unique disciplines. It also demonstrates the link between engineering science and the construction process. There will be far more difficulty in constructing a large cavern than a small tunnel to precise specifications.

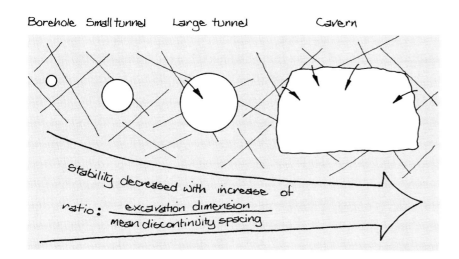

It is also worth noting how the rock mass behaviour changes with the ratio of excavation dimension to discontinuity spacing. From intact rock properties (which may be treated as being homogeneous and isotropic) the rock mass becomes highly inhomogeneous and anisotropic until, with relatively frequent discontinuities, it again becomes more homogeneous and isotropic, like a soil.

2.4.2 Depth and stability

There are two main modes of failure around underground excavations. One is block failure, where pre-existing blocks in the roof and side walls become free to move because the excavation has been made. The other is where the stress conditions induced in creating the excavation reach the local strength of the rock mass.

Nearer the surface, the stress is low and the discontinuity frequency tends to be higher. Therefore, block failure tends to be prevalent in near-surface excavations. Deeper in the ground, where the fracture spacing is reduced and the natural stresses are high, failure is usually governed by the high stress levels, enhanced by the presence of discontinuities.

There are no universal standard solutions for design, because each solution is specific to the circumstances (scale, depth, presence of water, etc). Nonetheless, recognising and applying the underlying rock mechanics principles will help to optimise the design. The design of excavation and support systems for rock, although based on scientific principles, has to respond to practical requirements, for example in allowing for the disproportionate economic effect of scale as the space to be excavated becomes larger and more cost effective.

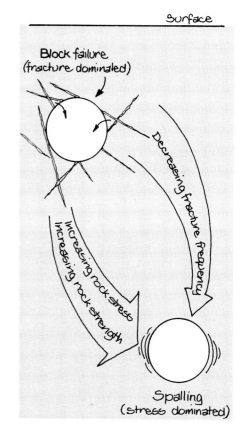

Key references

1. BROWN, E.T. [ed.] (1981) Rock characterization, testing and monitoring: ISRM suggested methods. Pergamon Press (Oxford)
2. GOODMAN, R. E. (1980) Introduction to rock mechanics. John Wiley and Sons (New York)
3. HOEK, E. and BROWN, E.T. (1980) Underground excavations in rock. Institution of Mining and Metallurgy, London

2.5 Permeability of rock

To predict the water flows into and around surface and underground excavations, the permeability of the rock must be established. For geothermal energy and radioactive waste disposal projects, the rock's permeability may be the single most important engineering parameter. Water inflows to a tunnel under construction may determine the tunnelling method. Generally, the fractures in a rock mass govern its permeability, but in some cases (e.g. sandstone) the intact rock may itself be relatively permeable.

The word permeability, used on its own, is a constant for a porous material (independent of the permeating fluid); its symbol is K (upper case) and its dimensions are (length)^2. The fluid is not always fresh water; it may be saline water, oil, gas, steam or a grout. Because nearly all civil engineering applications involve fresh water, it is convenient to use another term, the coefficient of permeability; its symbol is k (lower case) and its dimensions are (length/time). It is sometimes called the coefficient of hydraulic conductivity. The relation between the two terms is $k = K\gamma_w\eta$ where γ_w is the density of the fluid and η is its viscosity.

For the laminar (non-turbulent) flow of water through a porous material, Darcy's law links the amount of flow to the pressure gradient during the flow, i.e. $q = kia$, where a is the cross-sectional area through which the flow is taking place and i is the hydraulic gradient (the difference in pressure per unit length of flow path). For long-term drainage or pumping, the fluid-holding and other flow characteristics of the ground are also relevant.

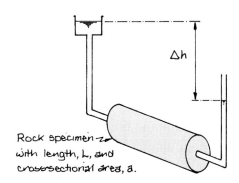

Rock specimen with length, L, and cross-sectional area, a.

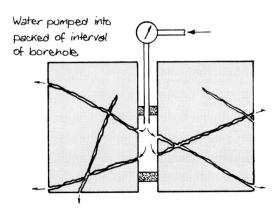

Water pumped into packed of interval of borehole.

Measurements to determine the coefficient of permeability can be on laboratory samples (usually of intact rock because of the difficulty of representing the fracture patterns) or by field tests at different scales. Tests in boreholes are influenced by discontinuities in the test region. Mass values of permeability are best determined by pumping tests. The very wide ranges in the coefficients of hydraulic conductivity for intact and fractured rock are shown below.

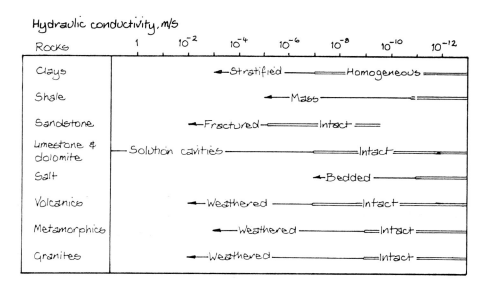

Hydraulic conductivity, m/s

When an excavation is made, the local permeability can increase, either because stress relief may open existing discontinuities, or because blasting may create new or wider fractures.

Flow has direction as well as magnitude (a vector); permeability is also directional, and the flow can take place within a three-dimensional network of discontinuities. Permeability should be considered as a tensor, like stress (see Section 2.3) in both continuous and discontinuous rock.

Water flow through discontinuities (rather than through the intact material) leads to large variations in the hydraulic conductivity values, depending on the sampling of the discontinuities. This is elegantly demonstrated by the work of Dr Jane Long (Key reference 4).

The effect of sample size and the connectivity of discontinuities within a sample can be seen in the diagrams below. The way that water might flow through the discontinuity systems is reflected by the 'rose' diagrams of permeability. The highly directional permeability of a single joint changes to more isotropic, all-round, stable values as more joints are included in the sample.

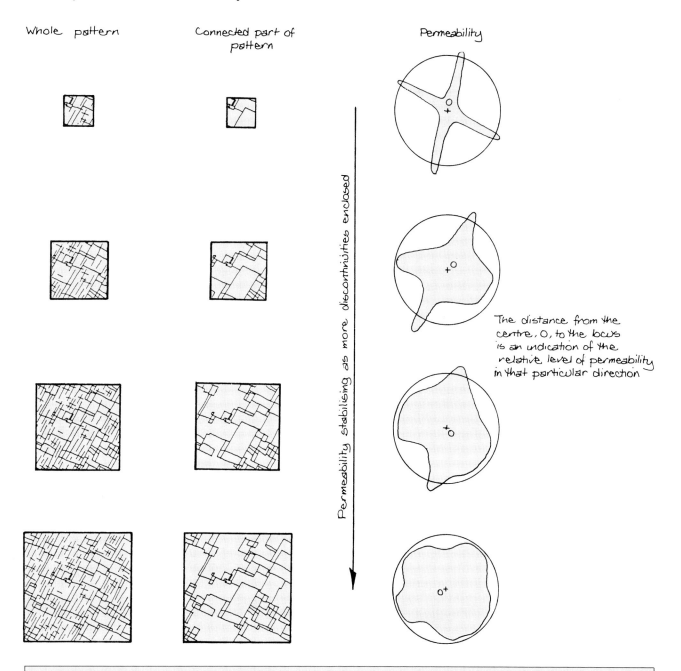

Whole pattern

Connected part of pattern

Permeability

Permeability stabilising as more discontinuities enclosed

The distance from the centre, O, to the locus is an indication of the relative level of permeability in that particular direction

Key references

1. HOEK, E. and BRAY, J. W. (1981) Rock slope engineering. Institution of Mining and Metallurgy, London
2. BRACE, W. F. (1980) Permeability of argillaceous and crystalline rocks. *Int. J. Rock Mech. Min. Sci.*, **17**, 241-251
3. ISHERWOOD, D. (1979) Geoscience data base handbook for modelling a nuclear waste repository, Vol. 1. NUREG/CR-0912 V1. UCRL-52719 V1.
4. LONG, J. C. S. Investigation of equivalent porous medium permeability in networks of discontinuous fractures. PhD dissertation, University of California, Berkeley

2.6 Anisotropy

The demonstration in Section 2.5 of the link between discontinuity patterns and permeability is a particular example of the concepts of anisotropy and inhomogeneity, which apply to all rock properties. Because of the way in which rock masses were originally formed, and because of their discontinuities, they may be expected to have different properties in different directions and at different locations. Rock is thus both anisotropic (the properties vary in different directions) and inhomogeneous (the properties vary at different locations).

A convenient way to represent the degree of anisotropy (in two dimensions) is by polar diagrams, in which the length of the radial arrow represents the value of that property in that direction. A circular polar diagram, i.e. a constant radius, represents two-dimensional isotropy. The anisotropy shown in the ellipse could apply to the deformability of the rock mass shown in the photograph, the joint sets causing the rock mass to be more deformable in the direction of the pointer, i.e. perpendicular to the joints.

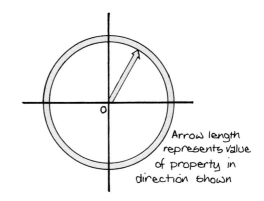

Arrow length represents value of property in direction shown

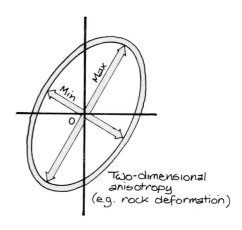

Two-dimensional anisotropy (e.g. rock deformation)

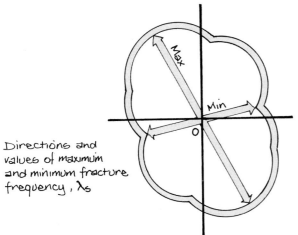

Directions and values of maximum and minimum fracture frequency, λ_s

Intact rock is also usually anisotropic because of its geological origin and history, but the anisotropy created by the discontinuities will have much greater engineering significance as, for example, with rocks such as shale and slate.

More complex anisotropy is caused by several joint sets. A good example is the frequency of discontinuities encountered in different directions through a rock mass. The distance from the centre to the locus indicates the number of fractures intersected per metre, λ_s, in any given direction, for a particular case. It is most important to note that the directions of maximum and minimum discontinuity frequency are generally not perpendicular.

This has important implications for design, because associated properties such as deformability and permeability exhibit similar trends, depending as they do largely on the discontinuity geometry.

Key references

1. ATTEWELL, P. B. and FARMER, I. W. (1976) *Principles of engineering geology*, Chapter 5: Preferred orientation, symmetry concepts and strength anisotropy of some rocks and clays. Chapman and Hall (London)
2. AMADEI, B. (1983) *Rock anisotropy and the theory of stress measurements.* Springer-Verlag (New York)

2.7 Inhomogeneity

Inhomogeneity means that properties are different at different locations.

The diagram shows a simulated tunnel face in one rock type, contoured for rock compressive strength. The strength is highly variable, and for many applications it would be inappropriate to assume an average value, e.g. when choosing a method for tunnel excavation.

Because a large number of tests are usually required to characterise inhomogeneity, it is often convenient to use rapid, low-cost index tests such as point load or Schmidt rebound hardness tests.

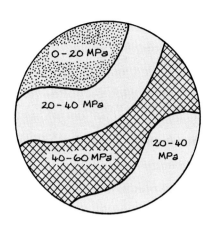

Inhomogeneity in rock properties can have either:

– a statistically homogeneous pattern, or
– no detectable pattern.

In the first case, the techniques of geostatistics are most useful. Geostatistics is not just statistics applied to geology, but the analysis of the variation in rock properties as a function of their location in space.

The diagram illustrates inhomogeneity of a rock property $P(x)$ along a line, such as a borehole, through a rock mass. This shows a wide range of values and there may be some trend in the variation. To identify this variation, the semivariogram approach of geostatistics can be used.

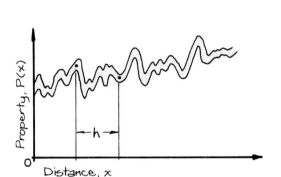

The values of the rock property along such a line are found at points separated by a given distance, h. We can then consider the variation along the line by studying the differences between these values. (This can also be done for additional lines and in two or three dimensions.) A statistical value $\gamma(h)$, related to the total variability between the values at points separated by distance, h, can be plotted against the distance, h, between sample points. This diagram is termed a semivariogram.

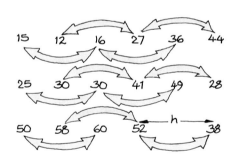

Sample pairs at distance h on a h/2 grid

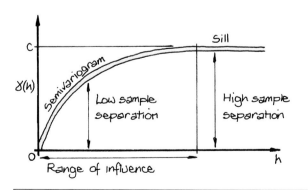

At very low separating distances, the semivariogram values will also be low. As h increases to beyond the range of influence, a, the values at a point located at 0 are no longer related to the values at points greater than a distance, a, away. At this distance, sample values are considered to be spatially independent of one another.

These geostatistical techniques are becoming increasingly useful in rock engineering because they indicate the region around a borehole that can be validly represented by the borehole test value.

Key references

1. CLARK, I. (1979) Practical geostatistics. Applied Science Publishers (London)
2. BROWN, E.T. [ed.] (1981) Rock characterization, testing and monitoring: ISRM suggested methods. Pergamon Press (Oxford)

2.8 Representative elemental volume

The effects of scale on rock properties, occurring because of rock inhomogeneity and fractures, cause test results on small-scale specimens of rock to be highly variable; this variability is increased by the presence of discontinuities. Indeed, one would not expect to obtain reproducible results for some of the properties (such as permeability) unless the tests were on a scale such that the test volume contained a sufficiently representative sample of discontinuities. The associated concept of a representative elemental volume (REV) is illustrated below. When the sample size is large enough that measured values are essentially consistent with repeated testing, the REV has been reached.

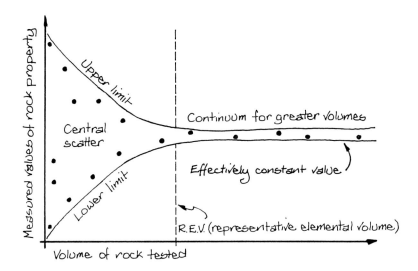

This concept applies to all rock properties that are affected by the discontinuity structure. The permeability of rock is a classic example; because of the very wide variation in hydraulic conductivity, together with difficulties in making adequate measurements, predicting the water flow in fractured rock can be a problem. For engineering purposes, it is usually sufficient to estimate inflows to an order of magnitude, e.g. to allow sufficient pumping capacity for the total inflows collected along a length of tunnel. Locally high inflows are hard to predict, although previous experience obviously helps.

Much more care is needed when analysing new engineering situations, because there are many factors yet to be researched. It is still questionable whether permeability is even a tensor, let alone a scalar, quantity, when water is flowing through an arbitrary network of discontinuities in a rock mass. For example, the directions of maximum and minimum permeability in a rock mass may not be perpendicular.

Other examples relevant to the REV concept are *in-situ* deformability and stress. In the latter case, stress is regarded as a point property and measured on a small scale within intact rock. However, to obtain reproducible measurements and eliminate the effects of stress concentrations caused by local discontinuities and larger geological features such as dykes, the stress would have to be measured over a very large rock volume.

The REV is thus a very important and powerful concept, which assists the initial design and interpretation of field measurements, and can even form the basis for engineering analysis – as is explained in Section 2.9.

2.9 Continuum and discontinuum methods of analysis

To model the rock mass for engineering purposes, we must consider whether it is a continuum or a discontinuum, i.e. whether it is a continuous material or one with significant interruptions in its mechanical integrity. Considered on a very small scale, soil is a discontinuum because it is composed of discrete grains. Relative to an engineering stucture, however, the individual grains are insignificant and the soil can be considered as a continuum. Techniques based on the theory of elasticity are then often appropriate.

If the fractures in a rock mass are fairly widely spaced, however, the rock block size can be similar to the engineering dimensions. In this case, it is not clear whether the rock should be modelled as a continuum or as a discontinuum. The REV concept (Section 2.8) is valuable for addressing this problem in the context of rock structure, rock stress and permeability. It is also helpful when establishing the basis for a theoretical model.

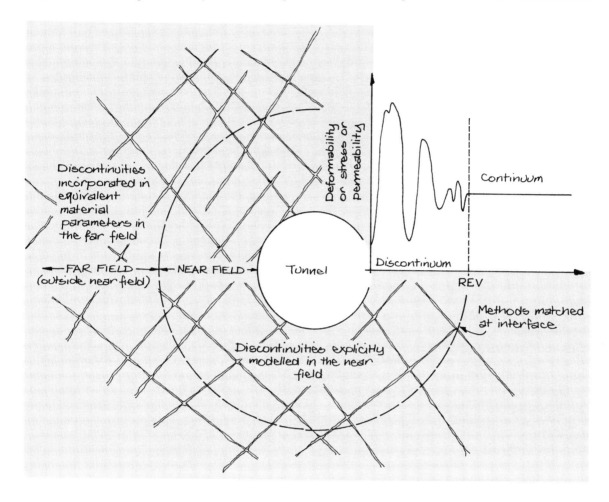

Note that the REV shown in the diagram could be for rock structure, *in-situ* stress or permeability, but would not be the same size in each case, and could be very large – say of the order of 10 m × 10 m × 10 m.

The current approach to modelling engineering projects in a rock mass is to treat the rock as a discontinuum (considering individual fractures) in the near field, and as a continuum in the far field (when the volumes are above the REV). An example of this is the analysis of stresses around a tunnel in a rock mass modelled by this hybrid technique.

Key references

1. LORIG, L. J. and BRADY, B. G. H. (1984) A hybrid computational scheme for excavation and support design in jointed rock media. ISRM Symposium, Design and Performance of Underground Excavations, British Geotechnical Society, Cambridge, 105–112
2. BROWN, E. T. [ed.] Analytical and computational methods in engineering rock mechanics. George Allen and Unwin (London) Chapter 4, Distinct element models of rock and soil structure, Cundall, P.

2.10   Interactions in rock engineering

For any rock engineering project, there will be certain primary factors to be taken into account in deciding on the correct engineering approach. In most cases the primary factors are rock mass structure, *in-situ* stress, water flow, and constructional aspects (which include cost, method and individual working operations). There may be additional primary factors for special projects (e.g. rock temperature for a geothermal energy project). Each of these primary rock factors has to be studied separately, but not in isolation, because during the engineering the factors interact with one another.

One method of identifying, studying and, indeed, thinking about how one factor affects another is the interaction matrix. This is an especially useful technique for new engineering projects where there is no precedent practice. The matrix can be presented at different levels of detail. The example below shows some interactions between the four main factors. The numbers in each box follow conventional matrix notation, i.e. the first number denotes the row and the second number denotes the column.

Through this mode of presentation, a rock engineering project can be studied in its entirety and in parallel, rather than separately and (possibly misleadingly) in series.

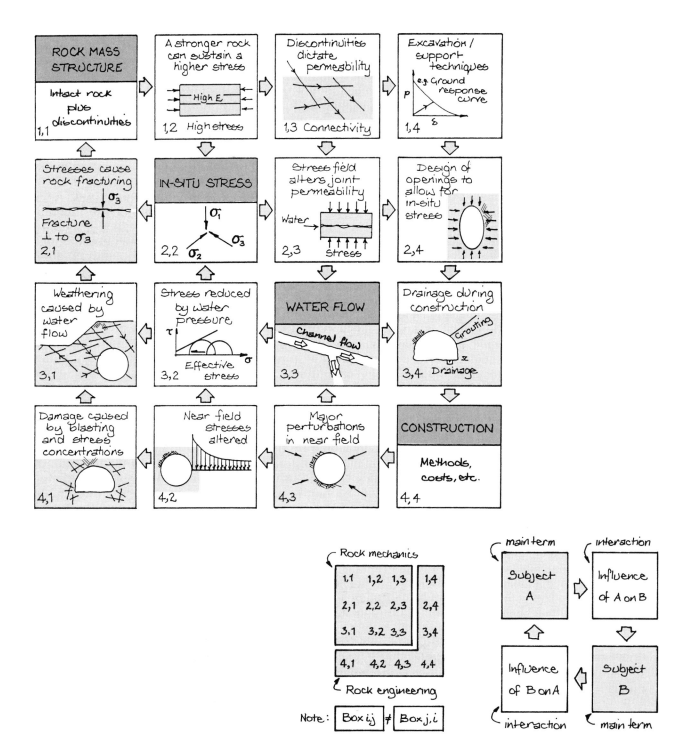

3

Measurement of rock behaviour and rock properties

Rock is natural not artificial; its properties have to be measured – they cannot be specified directly. Thus the measurement of rock properties is one of the most crucial aspects of rock engineering. Consider how much of the ground is sampled during a site investigation prior to a civil or mining engineering project. Is it 10%, 1%, 0.1%, 0.01%, 0.001%, 0.0001%, 0.00001%, 0.000001%, 0.0000001% ? The exact percentage will vary with the project but will be at the latter end of the scale. The problem which then naturally arises in site investigation is whether to obtain a few accurate results from a small number of 'fundamental' tests or a large number of inaccurate results from index tests – or a combination of both. An index test is usually relatively simple and rapid but does not provide a 'fundamental' property. If, however, an index test value can be correlated with a fundamental value and the bias understood, the test can be extremely useful. The point load test is a good example of the index test approach.

The measurement of rock behaviour and rock properties requires skill. There are a variety of scientific measurement systems available; the skill lies in choosing the optimal tests, test locations and number of tests for the specific project in hand.

Good engineering is multi-phased and constantly readjusted as new information is obtained. There is no definitive guidance on the total site investigation requirements for any rock engineering project, but the International Society for Rock Mechanics publishes Suggested Methods giving recommendations for specific testing methods (see Key reference 3, page 30).

A great deal can be established about a rock mass by careful visual observation, together with the answers to a series of key questions. How fractured is the rock? Is the rock weathered? Is the rock strong? Are the edges of the blocks broken off? Which are old fractures and which are new? What would happen if a block were removed? Is the rock wet? Has the rock debris on the floor deteriorated?

An enormous amount can be observed without even beginning the measurements. The skill lies in tailoring the measurements to the rock being investigated and the type of engineering project in hand.

3.1 Index tests

Rock property tests can give highly variable results, owing to inhomogeneity and the profound effect of discontinuities. Do we conduct tests of a fundamental nature (which as we have seen with the REV may require huge rock volumes) or do we conduct index tests?

Whatever we are measuring, and for whatever purpose, it is very important to apply the concepts of accuracy, precision, resolution and relevance.

Some typical index tests

Point load test   Sonic velocity test   Schmidt hammer test

Accuracy is the ability to obtain the correct answer on the average. In other words, there is no bias in the measuring technique.

Precision is the spread of repeated measurements, usually indicated by the standard deviation or variance. The spread of results is a separate consideration to whether or not the results are accurate.

Resolution is the number of significant figures to which measurements have been made. To observe the rock behaviour, good resolution may not be required; but to compare, say, measured displacement values with those obtained from numerical analysis, good resolution is necessary.

Relevance refers to whether the measured values assist in the fundamental objective of the site investigation or monitoring programme. **Because the objective varies with the project, there can be no standardised package of tests.**

Assessing index tests by applying the concepts of accuracy and precision is illustrated in the diagrams below using the analogy of firing range targets.

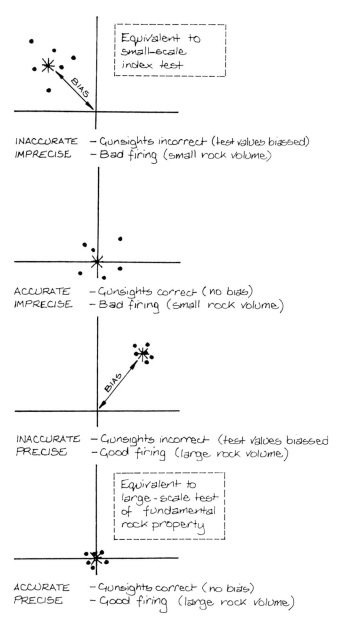

The small-scale index tests illustrated on the previous page are useful because they are rapid and cheap – and if the bias is known (see top diagram) the fundamental property can be estimated, as when estimating the compressive strength from the point load test value. A useful summary of index tests is given in Key reference 1 below.

The most popular index test is the point load test, which has been developed from indirect tensile strength tests. The rock is compressed between two conical platens until failure occurs. From the index value, I_s, determined from the distance between the platens and the load at failure, the compressive strength is estimated as about 24 $I_{s(50)}$ (where the I_s is standardised to a 50 mm dia. core specimen). A detailed description of this test is given in Key reference 2. Note that the value of 24 used as the multiplier may vary between about 18 and about 30.

Other index tests which have proved to be very useful are the Schmidt rebound hammer and sonic velocity tests, also illustrated on the previous page. These can provide index values in their own right or be used to estimate the compressive strength and elastic modulus. Because of the heterogeneous and fractured nature of the rock, many such index tests have been developed for a variety of purposes, e.g. for use in rock mass classification schemes, as described in Section 3.5.

Key references

1. FARMER, I. W. (1983) Engineering behaviour of rocks, 2nd ed. Chapman and Hall (London)
2. FRANKLIN, J. A. [co-ordinator] (1985) Suggested method for determining point load strength, International Society for Rock Mechanics, Commission for Testing Methods. *Int. J. Rock Mech. Min. Sci.*, 1985, **22** (2) 51–70
3. BROWN, E. T. [ed.] (1981) Rock characterization, testing and monitoring: ISRM suggested methods. Pergamon Press (Oxford)

3.2   Measurement of discontinuity characteristics

As emphasised throughout this book, the discontinuity characteristics of a rock mass are crucial. They can be broadly divided into geometrical and mechanical characteristics.

3.2.1 Geometrical characteristics

The table below indicates some of the more important geometrical discontinuity characteristics and how easily these can be measured from core, from videotapes of a borehole wall or downhole instrumentation, or on exposures where full access is possible.

Table 1   Measurement of geometrical discontinuity characteristics

Characteristic	Measurement method	Core	Borehole wall via TV camera	Exposure
Orientation	compass-clinometer	M	G	G
Spacing	measuring tape	G	G	G
Persistence	measuring tape	B	B	G/M
Roughness	against reference chart	M	B	G
Wall strength	Schmidt hammer	M	B	G
Aperture	scale or feeler gauge	B	M	G
Filling	visual	B	B	G
Seepage	timed observations	B	B/M	G
Number of sets	stereographic projection	M	G	G
Block size	3-d fracture frequency	B	B	G

G = good   M = medium   B = bad

These properties are illustrated in the diagram below and further described in Key references 2 and 3 (on page 33).

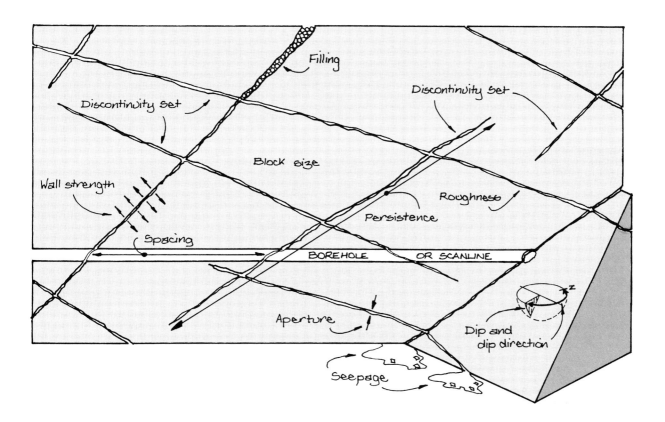

It is usual to summarise the intersection of discontinuities along a scanline by giving mean spacings or the frequency of occurrence, i.e.

Number of discontinuities $= N$
Length of scanline $= L$
Mean discontinuity spacing, $\bar{x} = L/N$
Discontinuity frequency, $\lambda = N/L$

Hence, $\bar{x} = \dfrac{1}{\lambda}$ and $\lambda = \dfrac{1}{\bar{x}}$

Discontinuities are not generally through-going for long distances in the rock mass although some, such as faults, may be. The idea of persistence is used in relation to the extent of discontinuities, and is generally characterised by the trace length. The trace lengths are important because they dictate the block size distribution of the rock mass. It is difficult, however, to estimate the three-dimensional characteristics from essentially two-dimensional trace lengths. Key reference 3 discusses this point.

The most widely used index of rock fracturing is the Rock Quality Designation (RQD). This is the percentage of core or scanline consisting of intact pieces longer than 0.1 m. The RQD can be estimated from the discontinuity frequency by the formula

$$\text{RQD} \simeq 100e^{-0.1\lambda}(0.1\lambda + 1)$$

This formula is derived using a negative exponential model for the discontinuity spacing histogram with a mean frequency of λ. Note that this formula estimates the RQD solely from the discontinuity frequency, which can be easily obtained.

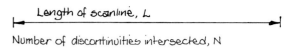
Length of scanline, L

Number of discontinuities intersected, N

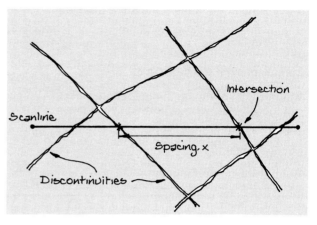

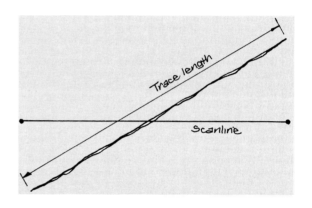

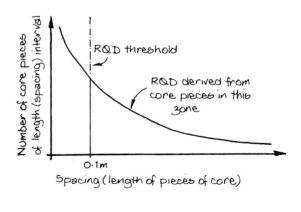

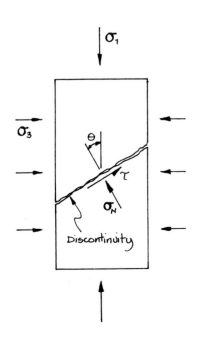

3.2.2 Mechanical characteristics

Some aspects of the mechanical characteristics of discontinuities have been mentioned in Sections 2.2 and 2.4. Probably the most important characteristics are those associated with failure of the discontinuity by slip, when particular combinations of normal stress and shear stress are applied to the discontinuity. The standard soil mechanics parameters of cohesion and angle of friction are used to characterise the failure mechanism of discontinuities. The possibilities are illustrated in the top diagram opposite, where normal stress is plotted on the x axis and shear stress is plotted on the y axis.

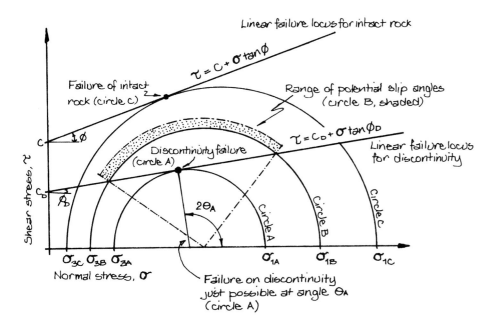

Linear failure locus for intact rock

$\tau = C + \sigma \tan\phi$

Failure of intact rock (circle C)

Range of potential slip angles (circle B, shaded)

$\tau = C_D + \sigma \tan\phi_D$

Linear failure locus for discontinuity

Discontinuity failure (circle A)

$2\theta_A$

Circle A   Circle B   Circle C

σ_{3C} σ_{3B} σ_{3A} σ_{1A} σ_{1B} σ_{1C}

Normal stress, σ

Failure on discontinuity just possible at angle θ_A (circle A)

Shear stress, τ

For a given set of principal stresses, the normal stress and shear stress can be represented by the conventional Mohr circle for any plane at angle θ to the maximum principal stress. Failure loci for intact rock and a discontinuity are shown on the diagram, with their respective cohesions and angles of friction. Generally, both parameters will be high for the intact rock.

In circle A, failure will occur along the discontinuity provided that it is orientated at exactly angle θ to σ_1. For a larger Mohr circle, B in the diagram, slip will occur along the discontinuity for the range of angles indicated. For a sufficiently large circle, C, which just touches the intact rock failure locus, the intact rock will fail by developing a failure plane at the specific angle indicated. This type of analysis can be applied to specific stability studies, but the strength parameters c and ϕ, determined for discontinuities, can also be used in the qualitative assessment of rock masses. More work on rock mass failure is required before we can measure c and ϕ correctly and understand their exact relation to field behaviour.

One way to measure the strength parameters c and ϕ is with the field shear box illustrated right. Typical shear stress displacement curves are also shown.

Another current research topic is the behaviour of discontinuities under different stress conditions before failure. The three plots of shear box tests show how the effect of increasing the normal stress changes the shape of the shear/stress displacement curve both before and after failure. This research is particularly directed towards assessing the influence of discontinuities on the deformability of the rock mass.

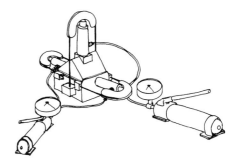

Portable shear box

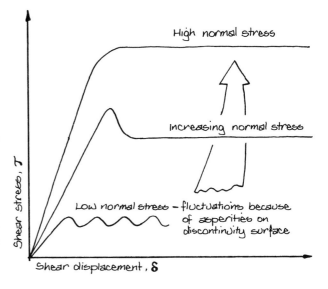

High normal stress

Increasing normal stress

Low normal stress – fluctuations because of asperities on discontinuity surface

Shear stress, τ

Shear displacement, δ

Key references

1. JAEGER, J. C. and COOK, N. G. W. (1979) Fundamentals of rock mechanics, 3rd ed. Chapman and Hall, (London)
2. BROWN, E. T. [ed.] (1981) Rock characterization, testing and monitoring: ISRM suggested methods. Pergamon Press (Oxford)
3. HUDSON, J. A. and PRIEST, S. D. Discontinuities in rock engineering (in preparation)
4. HOEK, E. and BRAY, J. W. (1981) Rock slope engineering. Institution of Mining and Metallurgy, London

3.3 Measurement of displacement

The simplest, most direct and most relevant measurement that can be made of rock behaviour, either in the laboratory or in the field, is the displacement between two points. To this day, the safety of many large structures such as dams and rock slopes is monitored on a regular basis by displacement measurement. This is because large-scale failure is usually preceded over a period of several days by a steady increase in the rate of displacement. In line with the discussion in Section 3.1, the measurement of displacement can be made easily, accurately, precisely and to a fine resolution, and is almost always relevant.

Displacement is a vector quantity. When the scaled components of displacement are expressed as longitudinal and shear strains, the resulting matrix of strains is a tensor quantity. Accordingly, care should be taken during measurement to ensure that sufficient displacement measurement is obtained – depending on the purpose of the measurements.

There is a variety of methods for measuring displacement, ranging from a measuring tape, through multi-rod borehole extensometers, to electronic techniques (see Key reference 1). There is also a variety of scales over which the displacement can be measured, from a few millimetres (when using a strain gauge), to many metres (using tapes) and many hundreds of metres (using electronic measuring devices).

One example of an accurate measuring device is illustrated in the diagram of a multi-rod extensometer below. A typical multiple point borehole extensometer record, from Dinorwig, is also shown (see Key reference 2).

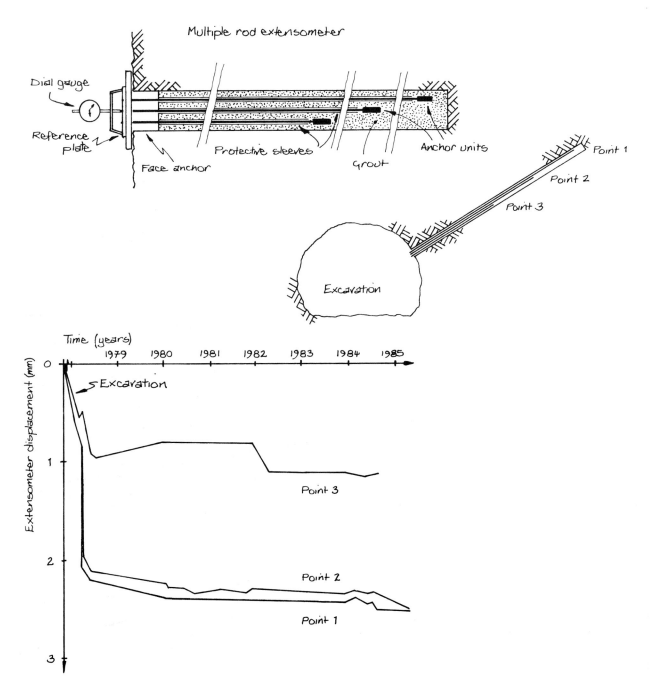

Kielder tunnel

The photograph left shows one of the drives of the Kielder experimental tunnel in the UK. In this tunnel, state-of-the-art displacement-measuring equipment was used to study the rock response to excavation and support, in three rock types and with a variety of support systems (see Key references 3 and 4).

The illustration below shows a photographic scan of a tunnel profile used for establishing the basic excavation geometry for the Dinorwig power station.

On the research side, displacement is probably the best parameter for verifying the validity of analytical techniques which model stresses and displacements in rocks. On the design side, displacement measurements can form the basis for support systems, as in the New Austrian Tunnelling Method, and more generally through the ground response curve described in Section 4.5 (see Key reference 5).

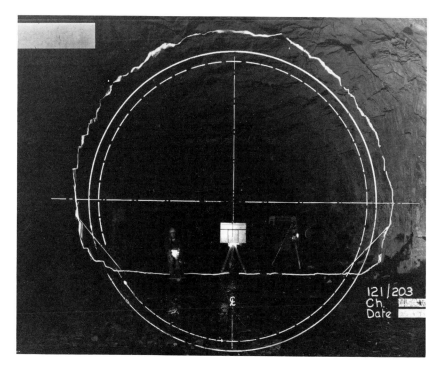

Tunnel measurement system

Key references

1. DUNNICLIFF, J. (1988) Geotechnical instrumentation for monitoring field performance. John Wiley and Sons (New York)
2. DOUGLAS, T. H., RICHARDS, L. R. and ARTHUR, L. J. (1983) Trial excavation for underground caverns at Dinorwig power station. *Geotechnique* **33** (4) 407–431
3. WARD, W. H., TEDD, P. and BERRY, N. S. M. (1983) The Kielder experimental tunnel: final results. *Geotechnique*, 1983, **33** (3) 275–291
4. WARD, W. H. (1978) 18th Rankine Lecture: Ground supports for tunnels in weak rocks. *Geotechnique*, **28** (2), 133–171.
5. JOHN, M. Arlberg expressway tunnel – Part 1 and Part 2 Investigations and design for the Arlberg expressway tunnel. *Tunnels and Tunnelling*, **12** (3), 46–51 and (4) 54–57

3.4 Measurement of stress

As described in Section 2.3, stress is a tensor quantity consisting of nine components, which can be listed in the matrix form shown. These components are with respect to an arbitrary set of x, y, z axes in three-dimensional space. The general trends of stress field versus depth are also described in Section 2.3.

Because the stress tensor has only six independent components, at least six separate measurements have to be made – whether these are by repeated testing (as with the flatjack) or simultaneously (as with the CSIRO gauge; see below).

Many stress measurement techniques exist. Four are recommended by the International Society for Rock Mechanics (Key reference 1). These are briefly explained below in the context of the stress tensor.

There are two vital factors to consider when measuring *in-situ* stress:

– the need for at least six separate measurements
– the fact that the measured stress in the near field may not be the same as that in the far field because of local alterations caused by excavation, including the measurement hole itself.

$$\begin{bmatrix} \sigma_{xx} & \tau_{xy} & \tau_{xz} \\ \tau_{yx} & \sigma_{yy} & \tau_{yz} \\ \tau_{zx} & \tau_{zy} & \sigma_{zz} \end{bmatrix}$$

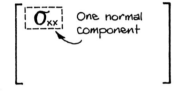

3.4.1. Flatjack method
(one-dimensional)

One normal component is obtained by measuring the distance between points P_1 and P_2, cutting a slot in the rock, inserting a flatjack and pressurising it until the distance between P_1 and P_2 is the same as before the slot was cut. Thus one normal component, say σ_{xx}, of the stress tensor is measured. Five more slots cut at appropriate orientations are tested to establish the complete stress tensor.

One normal stress component, σ_{xx}, is measured by this technique, but the shear stresses are not. It is necessary therefore to derive the stress components of the tensor from the six measured normal stresses. (This is directly analogous to the use of strain gauge rosettes.)

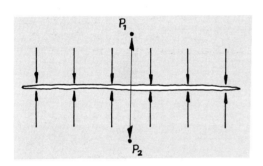

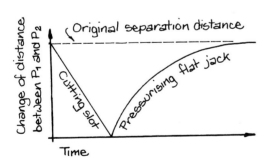

3.4.2 Hydraulic fracturing

A length of borehole at depth is isolated by inflatable packers and the test section pressurised with water until the rock fractures. From the values of peak (breakdown) pressure and the residual (shut-in) pressure, and with assumptions concerning the stress field (i.e. the principal stress directions), the three principal stresses can be estimated, as shown in the matrix.

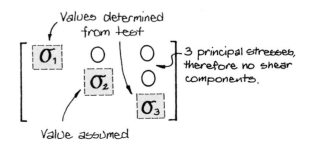

The hydraulic fracturing method has a major advantage over the other three methods described here: it can be used to estimate the stresses at very large distances from the observer, even to several kilometres away. The other methods are limited to a maximum of several hundred metres, and in practice are usually used a few tens of metres from the observer.

Because only two main readings are obtained in any one hydraulic fracturing measurement, assumptions have to be made regarding the stress tensor. These are that the borehole is parallel to a principal stress direction, and that the vertical stress can be estimated from the weight of the overlying material.

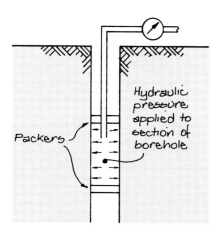

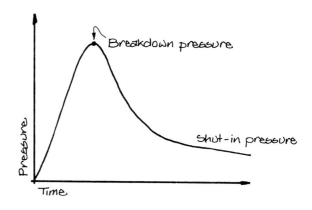

3.4.3 US Bureau of Mines (USBM) gauge
(two-dimensional)

The USBM gauge is a torpedo-type device which is placed in a borehole. The borehole is 'overcored' using a larger-diameter bit, so that the stresses around the original borehole are removed. The gauge measures the changes across three diameters of the borehole, thus providing three pieces of information. If the reference axes are aligned parallel and perpendicular to the borehole, the three measurements allow the three components of a two-dimensional stress field to be estimated, say σ_{xx}, τ_{xy}, σ_{yy} (see diagrams). Note, however, that to estimate the stresses from the three displacement measurements requires a knowledge of the stress/strain behaviour of the rock. This is assumed to be linearly elastic, and the rock is assumed to be continuous, homogeneous and isotropic. Moreover, to estimate the three-dimensional state of stress, the device should be used in three mutually perpendicular directions.

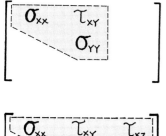

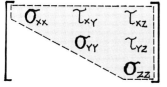

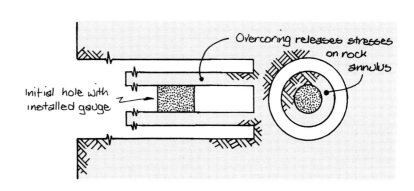

3.4.4. Commonwealth Scientific and Industrial Research Organisation (CSIRO) gauge
(three-dimensional)

Using the same operating principles as the USBM gauge, but with nine or twelve gauges, enough data are obtained to establish the complete three-dimensional stress tensor from one set of overcoring measurements (see stress matrix above). The CSIRO gauge, unlike the USBM torpedo, is glued into the borehole and is not reuseable.

Stress is a point property, and so measured values in a discontinuous rock mass should be expected to be highly variable. A wide scatter in measured stress values does not necessarily indicate bad measurement practice.

Key references

1. KIM, K. and FRANKLIN, J. A. [co-ordinators] (1987) Suggested methods for rock stress determination, International Society for Rock Mechanics, Commission on Testing Methods. *Int. J. Rock Mech. Min. Sci.*, **24** (1), 53–73
2. STEPHANSSON, O. [ed.] (1986) Rock stress and rock stress measurements. Proceedings of the International Symposium, Centex, Luleå.

3.5 Rock mass classification schemes

A rock mass classification scheme is analogous to a large but semi-qualitative index test, in which the rock mass is rated according to the values of a variety of input parameters. Such schemes are useful when they are applied to projects similar to those for which the classification techniques were derived, but can be very misleading when inappropriately applied – especially when there is no precedent practice. The two best-known schemes are those developed by Bieniawski and Barton (see Key references 1 and 2). Because of space limitations, only Bieniawski's scheme is outlined, to illustrate the general principles.

The problem with classifying rock masses is the apparent uniqueness of each rock formation, which makes generalisations inappropriate. In most major rock engineering projects, individual rock mass classification schemes are developed (or adapted from existing schemes) to suit particular geological conditions experienced during the progress of the works. These classification schemes are thus empirical and project-specific.

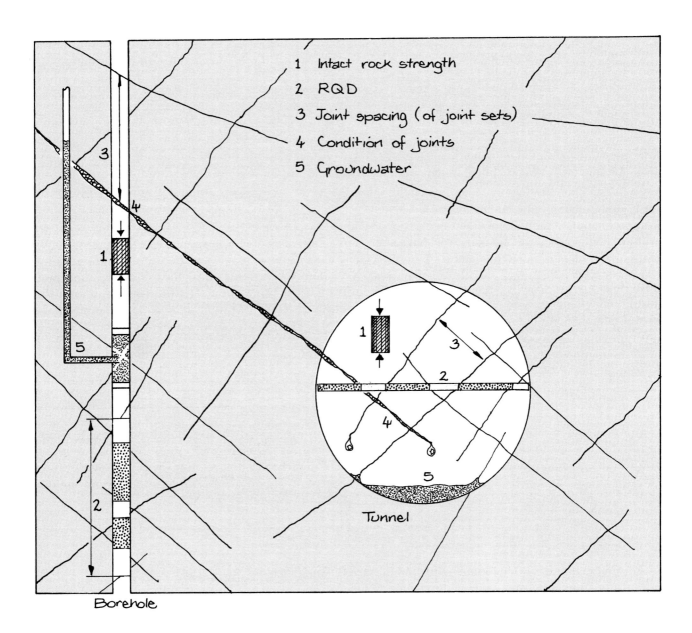

1 Intact rock strength
2 RQD
3 Joint spacing (of joint sets)
4 Condition of joints
5 Groundwater

Tunnel

Borehole

Values (or scores) assigned by Bienawski to the five parameters illustrated above are taken together to produce a single numerical value characterising the rock mass quality. The table opposite shows how this is done. Note that adjustments for joint orientation and other factors, not included here, are also needed.

In this way a classification rating, called the Rock Mass Rating (RMR), is produced, and the stability and support requirements of underground excavations can then be estimated. Such rock mass classification schemes can be very helpful in providing a guide for rock engineering design, not only for support but also for excavatability. However, it is to be expected that classification schemes will gradually be phased out, as rock mechanics principles are directly applied to the engineering problems in hand.

3.5.1 Geomechanical classification of jointed rock masses
(see Key references 1 and 2)

Table 2   Classification parameters and ratings

(1) Strength of intact rock					
Uniaxial compressive strength (MPa)	>200	100-200	50-100	25-50	10-1
Point load strength $I_{s(50)}$	>8	4-8	2-4	1-2	Too low
Rating	15	12	7	4	2-0
(2) Drill core quality					
RQD%	90-100	75-90	50-75	25-50	<25
Rating	20	17	13	8	3
(3) Joint spacing	>3m	1-3m	0.3-1m	50-300mm	<50mm
Rating	30	25	20	10	5
(4) Joint condition	Very rough surfaces No separation	Slightly rough surfaces Separation <1mm	Slightly rough surfaces Separation <1mm	Slickensided OR Gouge <5mm thick OR Joints open 1-5mm Continuous joints	Soft gouge >5mm thick OR Joints open >5mm Continuous joints
	Hard joint-wall rock Not continuous	Hard joint-wall rock	Soft joint-wall rock		
Rating	25	20	12	6	0
(5) Groundwater (other groundwater conditions given in Bieniawski's original version)					
General conditions	Completely dry	Completely dry	Moist – only interstitial water	Water under moderate pressure	Severe water problems
Rating	10	10	7	4	0

3.5.2 Rock mass classes and stability examples
(determined from total ratings (1) to (5) in Table 2)

Table 3   Rock mass classes and stability examples

Rating	100-81	80-61	60-41	40-21	<20
Class No.	I	II	III	IV	V
Description	Very good rock	Good rock	Fair rock	Poor rock	Very poor rock
Average stand-up time for	10 years	6 months	1 week	5 hours	10 minutes
unsupported span of excavation	5m span	4m span	3m span	1.5m span	0.5m span

An important postscript to this subject is that the correct mechanisms have to be identified. In this context one must expect the unexpected and think through all the possibilities: earthquakes, nearby engineering activity, rock/structure interface, effective stress zero? What is the weakest link in the structure? What will deteriorate first? The rock mass classification schemes help the engineer but they are certainly not a substitute for understanding the mechanics of the problem at hand.

Key references

1. BIENIAWSKI Z. T. (1984) Rock mechanics design in mining and tunnelling. Balkema (Rotterdam)
2. BARTON, N., LIEN, R. and LUNDE, J. (1974) Engineering classification of rock masses for the design of tunnel support. *Rock Mechanics*, 1974, **6** (4), 189–236
3. FARMER, I. W. (1983) Engineering behaviour of rocks, 2nd ed. Chapman and Hall (London)

4

Excavation and support

At its most fundamental level, rock excavation is achieved by increasing the number of fractures and thus changing the size distribution from rock blocks to rock debris; see diagram below.

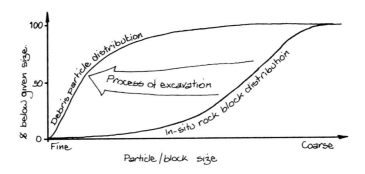

Therefore, on the stress/strain diagram illustrated, it is necessary to reach the post-failure zone in order to carry out the excavation. Once the excavation is complete, however, most of the rock surrounding it is in the pre-failure zone. Thus the rock defining the excavation is at the interface between the two major objectives of rock engineering, i.e. causing failure and avoiding failure respectively.

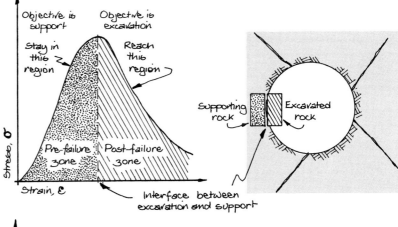

Rock is usually excavated by explosives (large rapid inputs of energy at periodic intervals), or by machines (lower rates of essentially continuous energy). These two extremes of energy are illustrated in the diagram on the right.

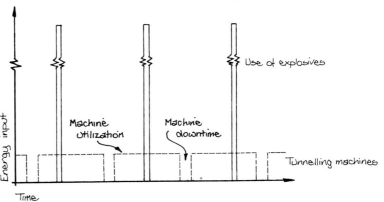

The properties of intact rock are important for excavation because the intact rock has to be broken. Conversely, the properties of discontinuities are important for stability because the discontinuities can allow failure to develop at low stress levels. There are cases where the natural discontinuity patterns make excavation and debris removal easier, e.g. when rock ripping.

On the next seven pages the most important aspects of excavation and support are described.

4.1 Blasting

Excavation by blasting involves the detonation of explosive placed in drill holes in the rock mass. Extremely high gas pressures are developed in the drill holes over a very short period, causing a dynamic stress wave to radiate out from the charge at sonic velocities. High gas pressures in the drill hole can be sustained over a relatively long period by continued burning of the explosives.

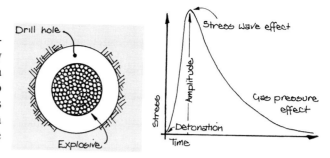

The result is compressive fracturing around the hole, and spalling caused by the compressive wave being reflected from a free face as a tensile wave. This is explained in the diagram and by the analogy of a train coupling failure.

Engine reverses into wagons.

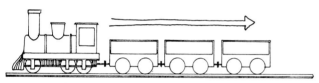

Compressive wave passes back through wagons.

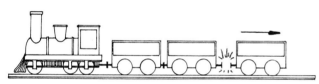

Last wagon breaks off if tensile failure of coupling.

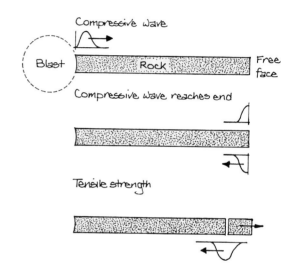

The presence of a free face for the tensile stress wave to develop is therefore one of the keys to successful blasting. For this reason most blasting rounds are designed on the basis of blasting to a free face, which either already exists or is progressively created.

The tensile strength is much less than the compressive strength, so although the rock can sustain a particular compressive stress wave, it may not be able to sustain a reflected tensile stress wave of comparable amplitude.

By the same token, discontinuities can cause problems, both from internal spalling due to reflection at dilated discontinuity surfaces, and from gas pressure effects.

Blasting always causes vibrations. The prediction of these and their effect on structures is clearly explained in Key references 3 and 4.

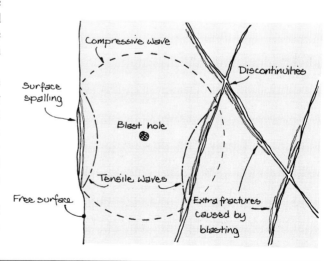

Key references

1. LANGEFORS, U and KIHLSTROM, B. (1973) The modern technique of rock blasting. John Wiley and Sons (New York)
2. HOEK, E. and BRAY, J. W. (1981) Rock slope engineering. Institution of Mining and Metallurgy, London
3. DOWDING, C. H. (1985) Blast vibration monitoring and control. Prentice Hall (New Jersey)
4. NEW, B. M. (1983) Ground vibration caused by civil engineering works. TRRL RR 53. Transport and Road Research Laboratory, Crowthorne, Berks

4.2 Pre-splitting

Pre-splitting is a blasting technique by which the final rock slope face is created first, such that subsequent bulk blasts do not damage the rock beyond the excavation boundary. The pre-split plane is created by simultaneous blasting in relatively closely-spaced, usually small-diameter holes, as illustrated in the diagram below.

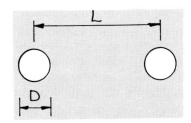

The most important keys to successful pre-split blasting are:

– drillhole accuracy : parallel holes
– drillhole diameter : must be small
– closely spaced holes (L/D<10)
– decoupled explosives (leaving an air gap between explosive and rock)
– simultaneous detonation

After successful pre-splitting, the 'half drillholes' can usually be seen, as shown in the photograph.

Parallel drillholes after pre-splitting

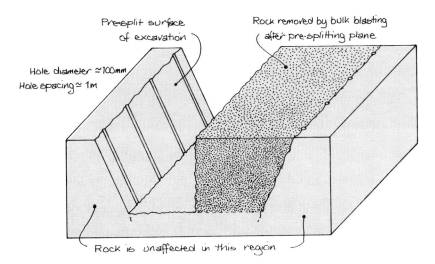

The pre-splitting technique is an excellent example of the successful application of rock engineering principles to achieve a specific objective – in this case a smooth final slope, with minimal damage to the rock behind and hence optimal safety (assuming that the slope is carefully designed – see Section 5.2), and minimal long term maintenance.

The technique is thoroughly described in Key reference 1. It can be used successfully in quite fractured and inhomogeneous rock, although it may be necessary to space the holes more closely. The presence of a high *in-situ* stress field could, however, inhibit development of a smooth plane, and smooth-wall blasting may be substituted.

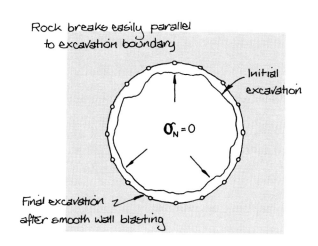

Deep underground, pre-splitting is ineffective because rock breaks perpendicular to the least principal stress. However, once an excavation is formed, one of the principal stresses is perpendicular to the excavation boundary. Smooth-wall blasting afterwards is then very effective; see diagram. For rock surfaces with weakened joints, or where the thickness of rock to be removed is relatively small, smooth-wall blasting can also be effective.

Key references

1. MATHESON, G. D. (1983) Presplit blasting for highway rock excavation. TRRL LR 1094, Transport and Road Research Laboratory, Crowthorne, Berks
2. LANGEFORS, U. and KIHLSTROM, B. (1973) The modern technique of rock blasting. John Wiley and Sons (New York)

4.3 Mechanised excavation

The capability of machines which cut and excavate rock has improved greatly in recent years. The basic mechanisms and machines are shown below.

4.3.1 Drag pick

The drag pick is essentially a tool for soft rock, i.e. with a compressive strength $\sigma_c \leq 80$ MPa (or less if the rock is massive or abrasive). Roadheaders are commonly used in the mining industry and are versatile for all types of openings.

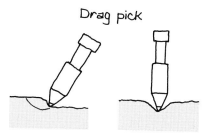

Drag pick

Disc cutter

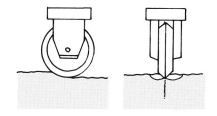

4.3.2 Disc cutters/button cutters

Disc cutters and button cutters may be used for the mechanical excavation of much harder rocks: up to $\sigma_c = 250$ MPa compressive strength with discs and using buttons for harder rocks. Full-face tunnel boring machines (TBMs) are best suited to long lengths of relatively homogeneous rock. Their operation can be prevented by very high strength rock (inability to cut) or by very weak rock (inability of reaction rams to operate). The example of a full face TBM is from the Kielder project.

Button cutter

Roadheader

Tunnel boring machine

4.3.3 Water-jet-assisted cutting

Water jets may be considered as an aid to mechanical cutting, because they decrease the required thrust and reduce tool wear.

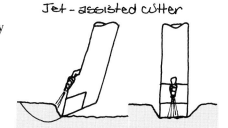

Jet-assisted cutter

Research is currently being conducted on the most appropriate ground parameters for predicting rock cuttability (e.g. fracture toughness, which is an alternative measure of strength, and discontinuity frequency). The nature of rock cutting with single and multiple tools needs further work in order that penetration rates can be increased, i.e. the rate at which excavation proceeds when the machine is operating. Optimal tool-cutting configurations also reduce vibration in the machine. However, the advance rate, i.e. the rate at which the tunnel is constructed, is related more closely to the system of tunnelling as a whole (i.e. including ground support, spoil removal, etc). The trend is towards more fully mechanised operations, with more and more excavation undertaken with tunnel-boring machines and water-jet-assisted roadheaders.

Key references

1. CIRIA Research Project RP351, Mechanical excavation of rock. Project Record available to CIRIA members only on request.
2. BENNETT, R. D. et al. (1985) State-of-the-art construction technology for deep tunnels and shafts in rock. Dept. of the Army, Technical Report GL- 85-1, Waterways Experiment Station, USA
3. NELSON, P. P. and O'ROURKE, T. (1983) Tunnel boring machine performance in sedimentary rock. Cornell University Press (New York)
4. PARKES, D. B. (1988) The performance of tunnel-boring machines in rock. CIRIA Special Publication 62

4.4 Block theory

For both underground and surface excavations, the relation between rock geometry and excavation geometry is essential in assessing the economics of both excavatability and potential overbreak.

There are two basic mechanisms of failure – structural failure, i.e. caused by discontinuities, and stress failure caused by high stresses.

Block theory is concerned with the three-dimensional configuration of rock blocks as determined by the discontinuity geometry, and how the stability of these blocks is affected by excavation – see box 4,1 of the interaction matrix in Section 2.10.

In block theory, the three-dimensional orientation of the discontinuities is used to assess the rock mass structure. When this structure is excavated, leaving new rock surfaces, the theory enables the instability of the exposed blocks and, indeed, the consequential instability of all the remaining blocks to be evaluated. The instability is a function of both the geometry and the mechanical properties. The block must be free to fall out – both geometrically and mechanically. As a result, rock reinforcement and support schemes can be coherently designed, from a knowledge of the mechanics of the potential rock mass failure.

Of course, the reliability of this analysis is a function of the validity of the rock discontinuity properties. Therefore considerable emphasis is now being placed on developing methods for analysing the three-dimensional nature of discontinuities. In recent years, major advances have been made in block theory. The analysis is extremely helpful in studying the basic design of excavation and support requirements.

An example of a block being released at an excavated surface is shown below.

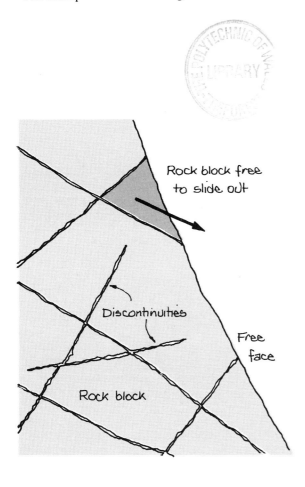

Block formation by planar discontinuities in the Bodmin granite, UK

Key references

1. PRIEST, S. D. (1985) Hemispherical projection methods in rock mechanics. George Allen and Unwin (London)
2. GOODMAN. R. E. and GEN H. S. (1985) Block theory and its application to rock engineering. Prentice-Hall (New Jersey)
3. WARBURTON, P. M. Implications of keystone action for rock bolt support and block theory. *Int. J. Rock Mech. Min. Sci.*, **24**(5), 283–290.

4.5   The ground response curve

The ground response curve is a general term for the movement of the rock resulting from excavation. The curve represents the force that would be required to hold the ground at various positions as inward displacement occurs.

When rock is excavated there are three main consequences:

– the rock is displaced, because resistance is removed by the excavation process;
– the normal stress perpendicular, and the shear stress parallel, to the excavation boundary are reduced to zero, the boundary becoming a principal stress plane;
– the hydraulic pressure in the excavation area is reduced effectively to zero and water flows towards the excavation.

These three primary effects are illustrated in the diagram. In connection with the ground response curve, the issue is the amount of rock displacement and whether support is required.

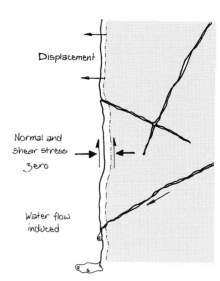

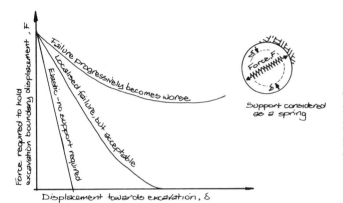

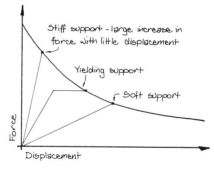

The ground response curve is a convenient way of determining the stability of the excavation and estimating the support requirements. The shape of the ground response curve depends on the rock and the excavation method. (Rock can be damaged by blasting, see interaction 1,4, page 28.)

The diagrams reflect two differing concepts. In the first case (above), three rock masses of different quality are compared: good rock associated with a particular excavation may need no support; reasonable quality rock may fail locally; poor rock fails progressively. In the second case (right), a single mass of rock is posited but three different excavation methods are adopted, such that their effect on the rock mass inhibits displacement at different force levels.

Thus the curve can be used for support design (see interaction 1,4, page 28) as in the observational approach of the New Austrian Tunnelling Method (Reference 5 on page 35).

There is a very important lesson to be drawn from the ground response curve. A good engineer does not attempt to stop all movement at an excavation boundary but, being aware of the movement, allows it to occur, and only controls further displacement where this is necessary. Not only would a large force (per unit of excavation area) have to be applied to prevent any displacement, but in applying it even more load could be attracted towards the support. In effect, therefore, support is an interactive process, matching the ground displacement and the yield of the support itself.

A final point is that the ground response curve may be damaged by the excavation process, and as a result the same rock could have different ground response curves depending on the excavation method, as shown in the diagram. Note from the support line that the result of bad blasting is to increase any support load necessary.

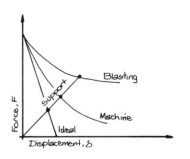

Key references

1. BROWN, E. T., BRAY, J. W. LADANYI, B. and HOEK, E. (1983) Characteristic line calculations for rock tunnels. *J. Geotech. Engng. Div.*, ASCE, 1983, **109** (1), 15–39
2. HOEK, E. and BROWN, E.T. (1980) Underground excavations in rock. Institution of Mining and Metallurgy, London

4.6   Rock reinforcement

If the rock fails because it is fractured and blocks can fall into the excavation, an elegant method of support is to reinforce the rock so that it behaves more like a continuum, as illustrated right.

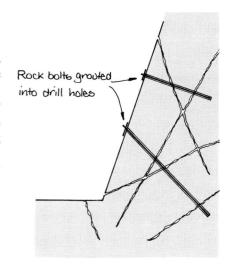

Rock bolts grouted into drill holes

To control slip on discontinuities, the normal stresses can be altered by bolting across the block interfaces. An important engineering decision is how to match the rock block structure with the rock bolt geometry. Bolting perpendicular to parallel discontinuities, along which there could be sliding, is perhaps the simplest example, but in practice it will be necessary to vary the angles of the bolts.

Reinforcement can take the form of rock dowels, cables, bolts or anchors. Bolting through discontinuities can have a dramatic effect on stability because, when pinned together, the rock behaves more like a continuum.

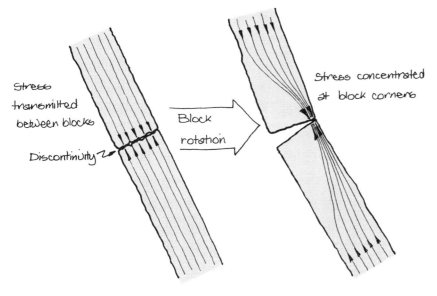

Stress transmitted between blocks

Discontinuity

Block rotation

Stress concentrated at block corners

It is not only slip but also block movement which is inhibited by reinforcement. Block movement can have severe effects, as illustrated left, because after rotation all the force formerly distributed along the discontinuity is concentrated at a block corner. Rock bolting and shotcreting inhibit rotation and thus reduce such stress concentrations.

The bolt may or may not be tensioned during installation, but even initially untensioned bolts develop tension because of rock movement.

There are many types of rock bolting methods; one of these is illustrated in the diagram left.

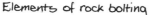

Elements of rock bolting

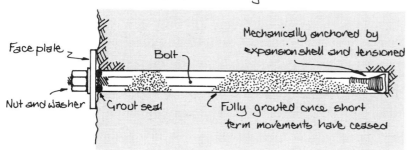

Face plate

Bolt

Mechanically anchored by expansion shell and tensioned

Nut and washer

Grout seal

Fully grouted once short term movements have ceased

Shotcreting or guniting (spraying rock with concrete or mortar with or without a mesh) can also be considered as methods of rock reinforcement. Their main effect is to help hold the rock together, by preserving the rock structure and maintaining the small rock blocks which lock the larger blocks in place. The shotcreting/guniting also prevents the rock drying out and minimises alteration to the joint infilling.

Key references

1. HOEK, E. and BROWN, E.T. (1980) Underground excavations in rock. Institution of Mining and Metallurgy, London
2. MAHAR, J. W., PARKER, H. W. and WUELLNER, W. W. (1975) Shotcrete practice in underground construction. US Department of Transportation Report FRA-OR&D 75-90
3. HANNA, T. H. (1980) Design and construction of ground anchors. CIRIA Report 65, 2nd ed
4. DOUGLAS, T.H. and ARTHUR, L.J. (1983) A guide to the use of rock reinforcement in underground excavations. CIRIA Report 101
5. BRITISH STANDARDS INSTITUTION, Draft for development, Recommendations for ground anchorages. DD 81: 1982, BSI, London

4.7 Rock support

If the ground response curve indicates that support is needed, and stability cannot be achieved by reinforcement, internal support has to be provided at the excavation boundary to stop further displacements.

Many types of direct support are available, from the occasional prop or arch, used for minor instability in mining, to the massive pre-cast concrete segments used for the Channel Tunnel. Tunnel linings (permanent supports for civil engineering) have been comprehensively reviewed by Craig and Muir Wood, Key reference 1. An example of support using arches is shown on page 35.

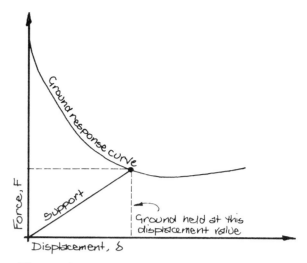

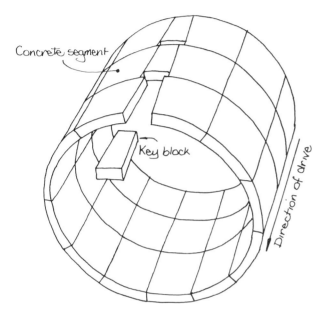

The provision of temporary support has a long history in mining and tunnelling, and continues to be based on judgement and experience. Wooden roof props are still used in coal mines, while steel arches in mine roadways are allowed to deform almost to buckling point as the roof, sides and floor slowly converge, before they are replaced. This body of experience has been encapsulated for specific conditions in various rock mass classification schemes (see Section 3.5), where a rock mass rating factor can be related to an indication of the stand-up times of unsupported excavations, or to the appropriate type of temporary support.

Bolted spheroidal graphite iron lining, British Rail Liverpool Loop

Key references

1. CRAIG, R. N. and MUIR WOOD, A. M. (1978) A review of tunnel lining practice in the United Kingdom. TRRL Report SR 335, Transport and Road Research Laboratory, Crowthorne, Berks
2. BRADY, B. H. G. and BROWN, E. T. (1985) Rock mechanics for underground mining. George Allen and Unwin (London)
3. O'ROURKE, T. D. [ed.] Guidelines for tunnel lining design. Prepared by the Technical Committee on Tunnel Lining Design of the Underground Technology Research Council, USA

5

Established applications

This section discusses established forms of rock engineering, and demonstrates the relevance of rock mechanics principles to these projects. New types of rock engineering are examined in Section 6.

There are some general principles that underlie both established and new applications. Design and construction techniques should match the excavation method and the structure to the ground conditions. There is no universal 'best method' of rock engineering, nor a perfect excavated structure; much depends on the design objective and the ground conditions. Under some circumstances, limiting rock displacements may be the primary concern, e.g. the Channel Tunnel. Under other circumstances, maximising rock displacements may be the objective, e.g. the block caving method of mining.

For design purposes, always check the history of similar projects from literature and other sources. Precedent practice (i.e. practice based on established methods) is invaluable experience and, through modern information technology, is more accessible. It is important to consider exactly which data are required from the site investigation programme, and to obtain as much information as possible from any adjacent works. Furthermore, it cannot be emphasised enough that there should be as much co-ordination as possible in the use made of this information, between site investigation, design, and construction.

As an example, one instance of precedent practice might have violated one of the principles, by using the stiffest form of support instead of the softest (see Section 4.5 on the ground response curve). The design with stiff support might have achieved the engineering objective, but at unnecessary cost. Modifying this precedent practice in the light of the principles described in Section 2 then integrates the two approaches.

Another example could be the existence of different ground conditions at a new site from those at a previous site where a particular design was successful. At the new site a different mode of failure might be possible, which a perceptive engineer would discover but one relying on precedent practice would not.

This is not to suggest that rock engineering should be practised directly from books, or by those without wide experience. Any rock project should be designed, and its construction supervised, by engineers experienced in working with rock and, preferably, with direct experience of similar types of project. But in several senses this begs the questions – of each rock mass being different (and so a new experience), of new types of project (so no-one has sufficient experience), and of how all engineers, at all levels of experience, have to learn as they take up new challenges. Whether the work involves predicting what will happen, which is design, or recording what has happened, which is experience, or responding to what is happening, which is the test of understanding, the engineer has to set the circumstances in a framework of basic principles, in order to understand the mechanics, to learn from experience, and to extend the principles. Design is an integral process, and there should be one person who has an understanding of the process as a whole. There are obvious pitfalls if there is a lack of communication at one of the design interfaces.

Nowadays, for large construction jobs, a trial excavation is used to study rock behaviour; this is the nearest approach to matching theory with practice before the final construction.

Rock bolts in a shaft

5.1 Foundations

It is often thought that building on rock is not a problem, but there are cases where the foundation rock must be studied very carefully, either because the rock is weak or fractured or because of the sensitivity of the structure.

The flow chart represents the design process; if the bearing capacity is reached or the settlement anticipated is excessive, the design is modified.

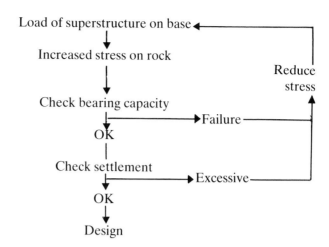

There is some difficulty in estimating the bearing capacity of rock. It is not practicable to test at full scale and, even at reduced scales, the representative elemental volume may be too large to test to failure. In practice there are two approaches: either to use a presumed bearing value, which is often given in local building codes for different types of rock, perhaps linked to empirical small-scale testing; or to establish deformability moduli relevant to the applied stress range (e.g. by the use of load bearing tests, see Section 3.1), to ascertain that the degree of settlement will be acceptable.

For major structures, the design approach should link the ground and superstructure movements, but this often requires numerical methods of analysis, which need to take into account the effects of discontinuities. The diagrams overleaf model contours of the same value of normal stress under line loads on transversely isotropic rock: they show the significant effect of the discontinuities on the stress distribution (see Key reference 2).

The difference between the stress distribution and the simple pressure bulb in isotropic rock can have several implications:

- the zone of settlement may be deeper because the rock is stressed to greater depth;
- the settlement will be asymmetrical if the discontinuities are inclined;
- overstressing beyond and below the edge of the foundation may occur, possibly even causing local shear failure.

Excavation for a nuclear reactor foundation, Torness, UK

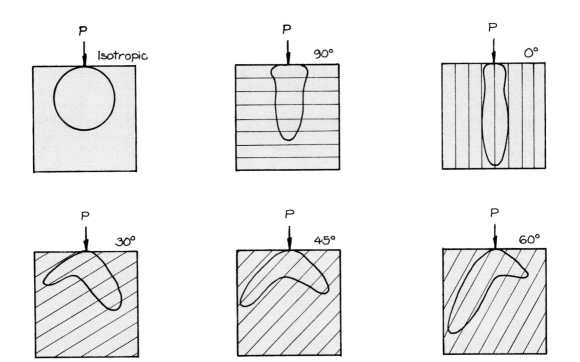

The stress distributions above also illustrate some other major difficulties encountered in rock mechanics and rock engineering. The pressure bulb in the top left diagram is only valid for an isotropic, linearly-elastic rock with uniform material constants, Young's modulus E, and Poisson's ratio ν. In the broadest terms, an elastic rock mass can have a maximum of 21 independent elastic constants. These are reduced to five independent elastic constants for the transversely isotropic rock mass of the other five diagrams. But how does one establish whether either the isotropic or the transversely isotropic assumptions have any validity or, indeed, any engineering usefulness? Efforts are now being made, in response to both questions, to model the rock more realistically and to obtain relevant rock characteristics. Also being studied is the question of how the actual loads, when excavated, compare with the design load.

The photograph below of the preparation of a rock surface for bridge piers shows the interaction between rock structure and foundation geometry.

Killicrankie pass, UK

Key references

1. GOODMAN, R. E. (1980) Introduction to rock mechanics. John Wiley and Sons (New York)
2. GAZIER, E. S. and ERLIKMANN, S. A. Stresses and strains in anisotropic rock foundations (model studies). Proc. Int. Symp. Rock Fracture, Nancy, Paper 2-1

5.2   Surface excavations and slope stability

5.2.1   Potential instability

A valuable method of considering the potential for slope instability (where failure is induced by pre-existing discontinuities) is to place instability overlays on a stereographic projection of the orientation of discontinuities (see Section 2.2). This enables the possibility of plane, wedge and toppling failure of rock blocks to be studied. The respective orientations of the discontinuities and the slopes, and the friction angle (described in Section 3.2), are taken into account.

The power of this method lies in the fact that the potential for failure can be quickly checked, as can the sensitivity of the solution to the input parameters. Thus the need for more rigorous analysis is swiftly established. The method also provides a graphical background for interpretation and a framework for site observation. The analysis can be computerised to display the stereograms and to rotate the instability overlays. Key references 1 – 3 give further details about the evolution of these methods.

Plane, wedge and toppling failure are the three major failure modes. Flexural toppling is also possible, and is described in Key reference 2. The diagrams below illustrate these modes of failure.

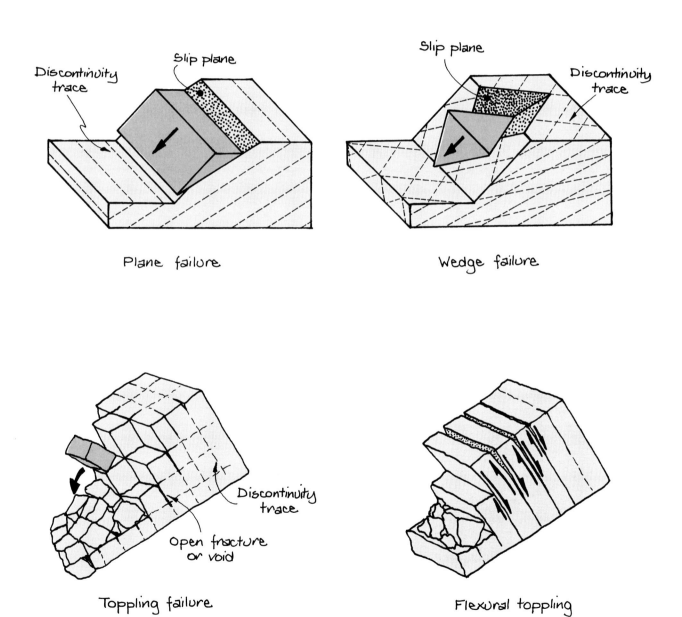

Plane failure

Wedge failure

Toppling failure

Flexural toppling

Using the stereographic projection: wedge failure

To study the potential for wedge failure, an overlay is placed over a stereographic projection showing the intersection lines between discontinuity sets. A wedge can fail if:

- the line of intersection of the planes at the bottom of the wedge daylights on the slope (so slip is geometrically possible), and
- this line plunges at an angle greater than the friction angle (so slip is mechanically possible).

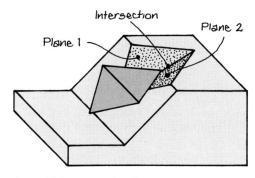

1. Initial slope design

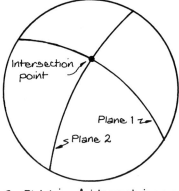

2. Plot line of intersection on stereographic projection.

These conditions are met if the shaded area of the overlay lies above an intersection line (plotted as a point on the stereoplot).

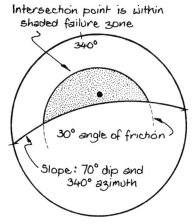

3. Use overlay for design parameters

For example, if the slope dips at 70° in a direction 340° (given the two planes in the diagram and a 30° friction angle), failure is possible.

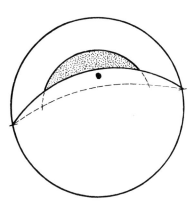

The unknown factor in the design is the slope dip angle for safety. By moving the great circle outwards until the intersection line is no longer beneath the overlay, the slope angle is made flatter and safe.

4. Adjust angle of slope to avoid wedge failure

Wedge failure in pre-split rock slope near Loch Lomond, UK

Similar techniques are used for plane and toppling failure, although the overlay geometries are different.

This method is especially valuable for considering all the slopes around a surface excavation. However, it indicates only the potential for failure, and further analysis is required following the methods outlined in Key references 1, 5 and 6.

Stereonet techniques can identify potential instability and the likely failure mode, and as such are first pass methods. The forces needed to restore stability depend on the geometry of the unstable rock mass, which is determined by the location and persistence of discontinuities. Neither of these can be taken into account in the stereogram, unless the site is divided into regions and the discontinuities graded according to persistence.

5.2.2 Design of surface excavation layouts

The layout (in plan) is only one of many factors to be considered in the design of a surface excavation. The stereographic methods outlined in the previous section can be used to study the primary modes of slope failure caused by pre-existing discontinuities, taking slopes at all dip directions. There will be some directions in which failure is far more likely than others, and so a circular plan for a surface excavation is most unlikely to be the optimal shape, because it takes no account of the rock structure. Conversely, if a circular excavation is required, the safe slopes can be found for all dip directions.

Open-pit copper mine

A judicious choice of excavation orientation can often have a dramatic effect on the potential for block failure, particularly in slope stability. For example, changing the direction of a highway cutting by, say, 30° in compass bearing could make all the difference between major instability problems and a totally safe slope. Similarly, for large surface excavations, because the excavated slope faces through 360°, an elliptical (or other) plan shape will almost always be more suitable than a circular one. The aim is to take advantage of the slope directions where failure is unlikely (or less likely). Using the overlay technique, the potential for plane, wedge and toppling failure in all directions can be studied. There is an excellent example of this in Key reference 2. Many excavations in rock are instrumented in order to study rock movement, as shown in the photograph.

It should not be forgotten that in weak, highly-fractured rock the failure mode can be a rotational slip like that in soil. Note too that blasting can introduce new discontinuities, and subsequent stability may differ from that estimated from the pre-blasting rock characteristics.

Measurement of displacement within a tension crack behind a slope failure

Key references

1. HOEK, E. and BRAY, J. W. (1981) Rock slope engineering. Institution of Mining and Metallurgy, London
2. GOODMAN, R. E. (1980) Introduction to rock mechanics. John Wiley and Sons (New York)
3. MATHESON, G. D. (1983) Rock stability assessment in preliminary site investigations – graphical methods. TRRL LR 1039, Transport and Road Research Laboratory, Crowthorne, Berks
4. MATHESON, G. D. (1986) Design and excavation of stable slopes in hard rock, with particular reference to pre-split blasting. In Rock engineering and excavation in an urban environment. Institution of Mining and Metallurgy, London, 271-83 517-24
5. WALTON, G. (1988) Handbook on the hydrogeology and stability of excavated slopes in quarries, HMSO, London
6. WALTON, G. (1988) Technical review of the stability and hydrogeology of mineral workings. HMSO, London

5.3 Underground excavations

5.3.1 Influence of depth and excavation dimensions

The dimensions and depth of an excavation strongly influence the mode of rock failure (see Section 2.4.1). In general, the rock structure dominates failure near the surface through block movement, whereas high stress concentrations induce failure at depth. We have also seen the effect of the relative scale of excavation dimension; the larger the excavation the more unstable it is.

The photographs on these two pages illustrate the potential for rock failure at different depths and for different sizes of excavation. Examples are given of three types of stress failure in a deep excavation (a gold mine).

Block failure potential in a small near-surface excavation

Highly persistent near-surface discontinuity in the Reigate Greensand, UK

Intersection of two joint sets influence the shape of the excavation roof

Block failure potential in a medium-depth large excavation

Large excavation for Dinorwig hydroelectric scheme in North Wales, UK

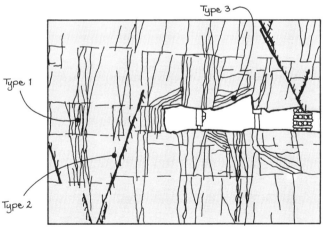

Stress failure in a deep excavation

The diagram on the right illustrates the extreme stress conditions at depth in the South African gold mines, and gives an indication of the fractures that can develop (see Key reference 5). The type 1 fractures are induced by the overall stress distribution; the type 2 fractures result from slip on pre-existing discontinuities; and the type 3 fractures are caused by local concentration of stress around specific tunnels and stopes.

Key references

1. HOEK, E. and BROWN, E.T. (1980) Underground excavations in rock. Institution of Mining and Metallurgy, London
2. BRADY, B. H. G. and BROWN, E. T. (1985) Rock mechanics for underground mining. George Allen and Unwin (London)
3. RICHARDSON, M. (1985) Geotechnical aspects of the Reigate sand mines. MSc Thesis, University of London
4. DOUGLAS, T. H., RICHARDS, L. R. and ARTHUR, L. J. (1983) Trial excavation for underground caverns at Dinorwig power station. *Geotechnique*, 1983, **33** (4), 407-431
5. Rockburst: prediction and control. (1983) Papers presented at a Symposium, Institution of Mining and Metallurgy, London

5.3.2 Underground block failure

Boreholes in rock are usually very stable because their diameters are small compared to the discontinuity spacing, as shown by the two boreholes illustrated in the photograph. In simple terms, there is not enough space for a block to fall into the borehole; in hard rock the borehole has to be at considerable depth before stress-induced failure occurs.

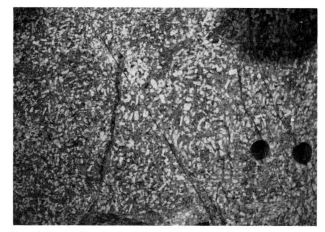

Boreholes in Carnmenellis granite, UK

As the excavation dimension increases relative to the discontinuity spacing (Section 2.4.1), however, there is much more opportunity for individual blocks to move. This is clearly demonstrated in the photograph on the previous page of the huge Dinorwig excavation, and can also be seen at intermediate scales in the photographs here.

Discontinuities forming unstable blocks in granite

Kielder roof support

In the Kielder experimental tunnel, there was repeated block failure of mudstone; small blocks fell from the excavated profile, ravelling back to two major, persistent discontinuity sets, one near-horizontal and one near-vertical. The photograph shows the support required to stabilise the roof of the tunnel, formed by the near-horizontal discontinuity set.

A simple geometrical construction based on the angle of friction alone is described in Key reference 1, and shows the potential for slip in underground excavations in layered rock. In these circumstances, block failure can be induced by the stress components acting along and across discontinuities, as an intermediate case between pure block failure and pure stress failure.

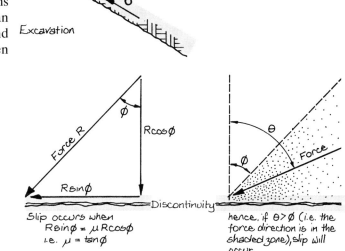

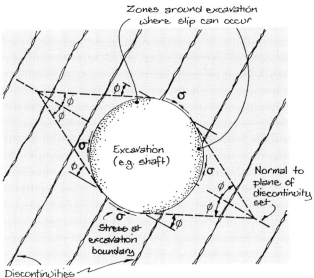

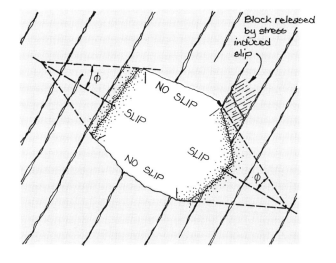

When an excavation is made, the principal stress component acts along the excavation boundary (see Section 4.5). Assuming the darker area shaded in the diagram above contains the resultant force, there is a potential for slip along the discontinuity, because the applied force is greater than the frictional resistance.

The geometrical construction which is used to find the regions around a shaft (or tunnel or, indeed, any underground excavation) where this type of slip could induce block movement is shown left.

The effect of a lower angle of friction (reduced for example by water flow) is to enlarge the region where slip is possible, as shown below.

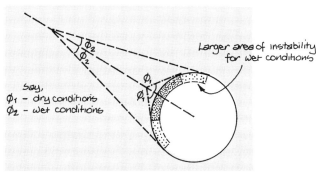

This simple technique can also be applied to irregular excavation shapes, as in the diagram left. Slip is not inevitable, however, because the stresses tend to become redistributed as weak zones develop.

For excavations (shafts, tunnels or caverns) at intermediate depths, the effects of stress can aid stability. The redistribution of stress generates circumferential stresses around the perimeter of the opening. The increase in normal stress across discontinuities provides greater frictional resistance to shearing. The stability of blocks in such zones can thus be increased. Because stresses increase with depth, the stabilising effect of the confining stresses also increases, and the requirement for reinforcing a particular block arrangement can decrease with depth (e.g. in a shaft). Conversely, rock blocks can be squeezed out at greater depths and, in more extreme cases, rock bursts can occur.

Key references

1. GOODMAN, R. E. (1980) Introduction to rock mechanics. John Wiley and Sons (New York)
2. WARD, W. H. (1978) 18th Rankine Lecture: Ground supports for tunnels in weak rocks. *Geotechnique*, 1978, **28** (2), 133-171.

5.3.3 Underground stress failure

The process of underground excavation can alter the stress field to such an extent that the rock fails. This failure is usually progressive because the rate of excavation is gradual but, if there is a major dynamic change in the potential energy state, a rock burst will occur.

The data required to assess the potential for stress-induced failure are:

– the strength of the intact rock;
– the stress distribution around the excavation.

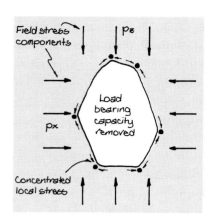

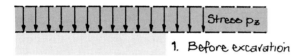

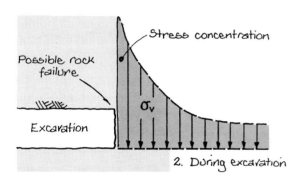

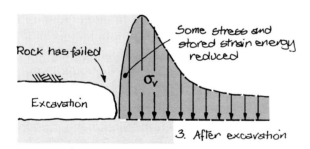

The significance of the potential for stress failure depends on the objective of the underground construction. For a civil engineering structure with a design life of, say, 100 years, block failure and stress failure have to be avoided in the long as well as the short term. Weathering and time-dependent effects will change the mechanical properties and stress distributions over the design life. When it is the rock itself which is required rather than the opening, as in mining, there is much greater flexibility in design, and particularly in the excavation sequence. This is illustrated by coal extraction methods: in the longwall system, the roof is planned to collapse behind the working face; the alternative of room-and-pillar mining relies on the roof and pillars remaining stable, at least until coal removal is completed. The deterioration of the Black Country limestone mines (see Key reference 4) is caused by time-dependent effects on room-and-pillar workings.

Large rock pillar in an underground excavation

Local stress failure need not necessarily be damaging to the excavation because it may lead to safer conditions, as in the case illustrated.

The South African mining example in Section 5.3.1 illustrates an extreme case of stress-induced failure.

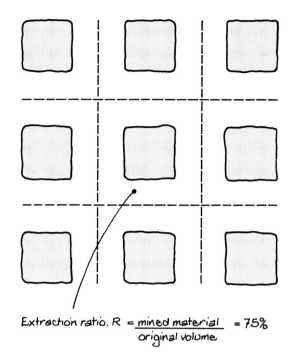

Extraction ratio, $R = \dfrac{mined\ material}{original\ volume} = 75\%$

With room-and-pillar support, as with all underground excavation where stress failure is possible, a factor of safety can be chosen in terms of the strength of the rock and the stress likely to be applied to it. The photograph on the previous page shows a typical rock pillar.

$$Factor\ of\ safety\ =\ \frac{Strength\ of\ rock}{Stress\ in\ rock}$$

The stress concentration factor expressed as a function of the extraction ratio, R, in room-and-pillar extraction is:

$$Stress\ Concentration\ Factor,\ SCF,\ =\ 1/(1\text{-}R)$$

where R is the proportion in plan of the material mined. This relation is illustrated in the graph.

Between 90% and 91% extraction ratio, the 1% increase in the proportion of mined material leads to a 10% increase in stress – and eventually to catastrophic failure, as R is further increased. The factor of safety is developed from the stress value, in conjunction with pillar strength.

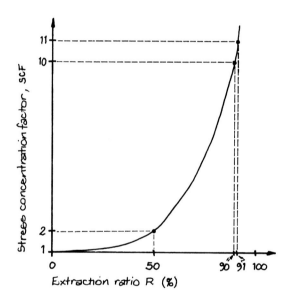

Apart from simple circular geometries and configurations such as the room-and-pillar system, there are very few situations where the stress distribution can be directly obtained, and it is then necessary to use one of a number of available computational methods of stress analysis. These include the finite element method (where the material is divided into an assembly of discrete interacting elements), the boundary element method (where stresses can be determined by the traction distributions on the boundary), the distinct element method (in which the discontinuous rock mass is modelled as an assembly of interactive quasi-rigid blocks), and hybrid methods combining two or all three of the methods described. These methods are developing rapidly, as hardware capacity increases and as programs are extended and refined. Key reference 5 contains a preliminary list of rock engineering computer programs compiled by the International Society for Rock Mechanics. What has not kept pace with software development is our ability to measure the material properties of the rock mass, or to establish the far-field *in-situ* stresses, or to account for the presence of water, or, most importantly, to verify the applicability of the models by full-scale studies.

Key references

1. HOEK, E. and BROWN, E.T. (1980) Underground excavations in rock. Institution of Mining and Metallurgy, London
2. PENG, S. S. (1985) Coal mine ground control, 2nd ed. John Wiley and Sons (New York)
3. Rockburst: prediction and control. (1983) Papers presented at a Symposium, Institution of Mining and Metallurgy, London
4. BROOK, D. (1988) Abandoned limestone mines in the West Midlands of England–underground dereliction in an urban area, in Legrand, M. (ed.) Cities and subsurface use. Balkema (Rotterdam)
5. ISRM Commission on Computer Programs (1988) Rock engineering software. *Int. J. Rock Mech. Min. Sci.*, **25**, (4), 1988

5.3.4 Underground mining

The major difference between rock engineering for mining and rock engineering for civil engineering projects is that, in mining, parts of the excavation state (surface or underground) will be temporary, and the design life required of these parts may be as little as a few days or weeks.

In cases such as longwall coal mining, the rock failure (caving of the roof following the face) is an integral part of the mining procedure. In other caving methods of mining, principally block caving, the rock mass is broken up using its own potential energy, as it is lowered towards draw points.

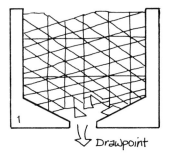
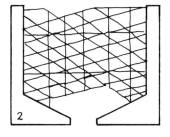
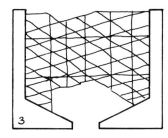

Drawpoint

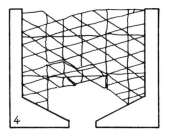
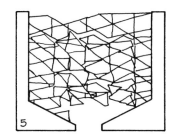
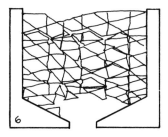

The block-caving system of mining can be used for very large volumes of rock. The diagram above (see Key reference 1) is a computer-generated model showing the principles involved. The potential energy of the rock is available initially to open up and induce new fractures. As blocks fail, the rock mass gradually loses its integrity, until blocks eventually flow through the draw point, as shown in the photograph.

The block size distribution which has to be achieved for flow to continue, and for convenient handling, is critical. As the rock mass moves downwards it dilates and, in some circumstances, forms a natural arch, supporting the mass above and inhibiting flow. Thus the geometry of the draw point has to be carefully designed, as does the sequencing of the rock removal. Optimising these factors depends on understanding and applying the appropriate rock mechanics principles.

Core sampling of an ore body illustrating pre-existing fractures

Ore being removed through a draw point

A component of the fragment size distribution retained in a 75mm sieve

The conservation of vertical load is a principle of stress distribution that applies to all forms of underground excavation, and so is a very useful concept.

Whatever shape or size of excavation is made, the vertical load that was originally supported by the rock is still present. This is taken up through the redistribution of stress as excavation occurs. In the diagram below (showing the theoretical stress concentration around a circular opening when there is only vertical loading) the loads represented by the two shaded areas are therefore equal.

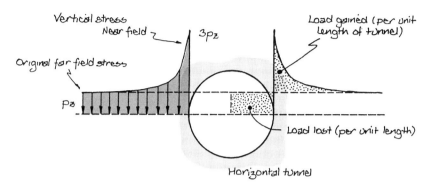

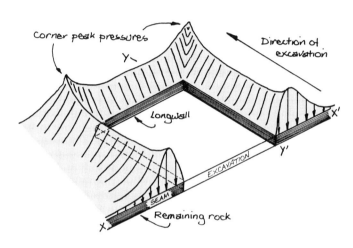

Another example of this principle is the stress distribution around a longwall face, the X-X section being similar to the diagram above, with stress concentrations on either side of the mined area. The principle must also hold for the Y-Y direction, but note that, at the sides of the excavation, the stresses have decreased.

The maximum stress concentrations are associated with the maximum excavation dimension, so that the high corner peak pressures are on the diagonal of the mined panel.

A natural extension of these ideas is the concept of a zone of influence – the region where the stress is altered by a specific percentage as a result of the removal of the rock. The 5% zone of influence is illustrated in the diagram below; beyond this zone the stress components are altered by less than 5% from the original *in-situ* component. This is valuable when considering adjacent underground excavations and is explained further in Key reference 2.

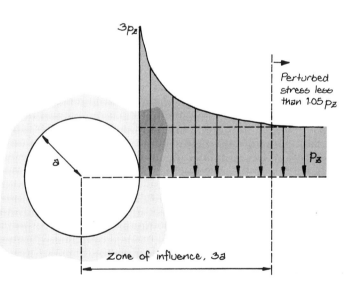

To conclude this section on 'conventional' applications, the following points are emphasised.

- A good rock engineer develops an understanding of rock mechanics principles.
- Traditional design methods are helpful because they embody formalised experience.
- Both an understanding of the principles and practical experience are needed in order to optimise existing methods, to be innovative in design, and to provide appropriate design for new projects in rock engineering.

Key references

1. VOEGELE, M. D. (1978) Rational design of tunnel supports: an interactive graphics based analysis of the support requirements of excavations in jointed rock masses. Final Report to the Department of the Army Contract No. DACW 45-74-C0066. PhD Thesis, University of Minneapolis, Minnesota, 1978, pages II-50 to II-54
2. BRADY, B. H. G. and BROWN, E. T. (1985) Rock mechanics for underground mining. George Allen and Unwin (London)
3. WILSON, A. H. (1983) The stability of underground workings in the soft rocks of the coal measures. *Int. J. Min. Eng.*, 1983, **1** (2), 91-187
4. WHITTAKER, B. N. (1975) An appraisal of strata control practice. *Min. Engr.*, 1975, **134** 9-24

5.4 Application example: tunnel portal design

The stages of a tunnel portal, a typical rock engineering project, are illustrated below, to show how, for this particular application, the principles of rock mechanics apply to all aspects of design and construction in rock. For each stage of the project, the aspects to be considered are listed below, with cross references to the relevant Sections. One of the preliminary stages is then examined further by considering the relevant interactions between the rock mechanics principles and the engineering operations.

5.4.1 The stages in portal construction

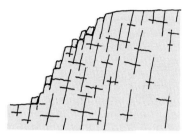

Stage 1 The portal site
- Geological context and groundwater conditions (Sections 1.2 and 2.5)
- Intact rock and rock mass characteristics (2.1-2.10, 3.1, 3.2)
- Rock mass classification schemes (3.5)

Stage 2 Surface excavation and temporary portal support
- Excavation and support (Section 4)
    Block and rock mass stability (4.4)
    Excavation methods (4.1, 4.2, 4.3)
    Temporary and permanent reinforcement (4.5, 4.6)
- Surface excavations and slope stability (5.2)
- Measurement of displacement (3.3)

Stage 3 Tunnelling
- Discontinuities, stress, deformability of rock masses and inflow (Sections 2.4, 2.9, 3.2)
- Index tests and measurement of stress and displacement (3.1, 3.3, 3.4)
- Excavation and support (4.1, 4.3, 4.7)
- Underground excavations (5.3)

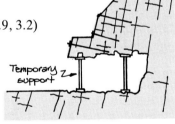

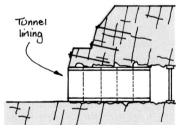

Stage 4 Lining
- Measurement of displacement (Section 3.3)
- Support
    Ground response curve (4.5)
    Rock reinforcement (4.6)
    Rock support (4.7)
- Underground excavation (5.3)

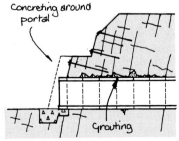

Stage 5 Permanent portal construction
- Excavation and support (Section 4)
- Foundations (5.1)
- Surface excavation and slope stability (5.2)

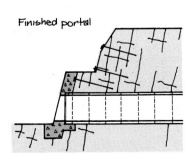

Stage 6 The finished portal
- Measurement of displacement (Section 3.3)

5.4.2 Interactions in rock engineering: stage 2 of the portal works

The first step in the design process is to take account of the interactions between rock characteristics, rock conditions, and construction operations; the quality of the final design depends on adequate planning in the initial stages. These interactions can be considered using the interaction matrix described in Section 2.10. Two examples are given below, one for excavation by pre-splitting and one for reinforcement of a rock face.

Example 1 links rock mass structure (Box 1,1) and the construction operation (Box 2,2), showing that each can affect the other.

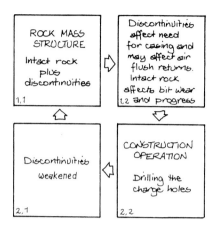

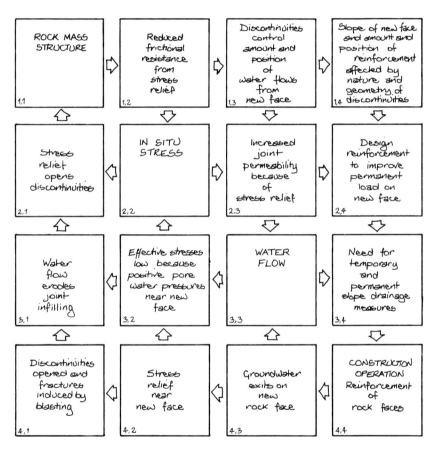

Design is itself an interactive process, however; in example 2, reinforcement of the face, the slope of the face and the reinforcement scheme are designed before excavation starts. As the geology is revealed and the actual discontinuities exposed, the design is reviewed to check the validity of the assumptions on which it was based. When necessary, the slope or the reinforcement pattern is amended. Many of the key references emphasise the need for this interactive approach to design.

When developing an interaction matrix, the following algorithm is useful, and has been applied to both the examples above. The starting point is the proposed construction operation (i.e. the bottom right hand corner of the matrix). Consider next the main diagonal of the matrix – the principal factors which will influence the construction. Finally, the process of evaluating the interaction of these factors will reveal the design parameters.

The two examples cited are operations in Stage 2 of the portal works. It is clear that the choice of the new slope of the work face will affect the blasting design and the choice of reinforcement.

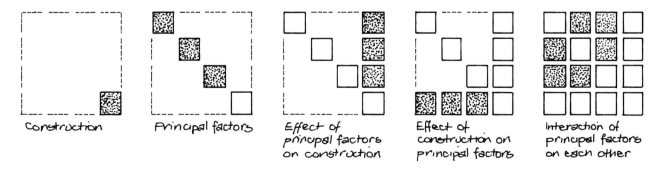

6

New applications

Rock engineering is now needed in order to solve problems that have never been solved before; for example, hot dry rock geothermal energy extraction and radioactive waste disposal. There is also a considerable number of other innovative applications being developed. These involve complex interactions, not only between rock structure, rock stress, water flow and construction, but also bringing in other primary factors, such as time and temperature. Because there is no previous experience on which to draw, an understanding of the principles is essential.

6.1   Geothermal energy

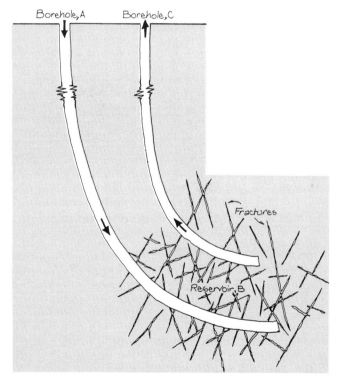

Hot dry rock geothermal energy is a method of extracting heat energy from hot rocks at depth. The concept is straightforward, but its implementation presents an interesting challenge to engineering.

The method adopted is deliberately to fracture further, at the base of suitable boreholes, the already naturally fractured rock. Water is pumped down borehole A, forced through the fissures in the hot dry rock reservoir B, to return to the surface up borehole C. Heat absorbed by the circulating water can then be extracted to provide energy.

Both concept and implementation depend critically on interactions between:

- Rock mass structure: the water flow depends mainly on the discontinuities.
- *In-situ* stress: most of the water flow will be along the widest discontinuities, which will tend to be perpendicular to the least principal stress.
- Water flow: the connectivity and other characteristics of discontinuities govern permeability, and therefore flow rates and head losses.
- Thermal properties: the heat balance between what is extracted by the water and what is drawn in from outside depends on the thermal conductivity of the fractured rock and the surrounding rock.
- Extraction rate: the system must be adjusted so that the rate of heat flowing into the reservoir from the surrounding rock is sufficient for steady state conditions to be achieved as heat is extracted.

An extended version of the interaction matrix shown in Section 2.10 may be created by including thermal properties and extraction rate as the fourth and fifth leading diagonal terms, with construction as the sixth.

Hot dry rock research station, Rosemanowes quarry. Cornwall, UK

The photograph above shows the UK hot dry rock research site at Rosemanowes quarry; that on the right shows the micro-seismic events generated by pumping water from one borehole, through the rock mass, to the other borehole. The locations of such events can be determined fairly accurately, and provide a wealth of information concerning the water flow process. For example, low-frequency emissions are thought to be associated with shear movement on discontinuities, while high-frequency emissions are thought to be associated with brittle failure, i.e. generation of new fractures. Note that the micro-seismic emissions are mainly below the boreholes, indicating considerable water loss and rock failure induced by water pressure.

As our understanding of the flow processes improves, so will the design of the production system. The potential energy generation from this heat source is enormous.

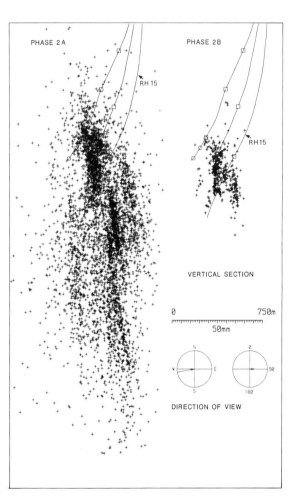

Micro-seismic events generated by pumping operation

Key references

1. PINE, R. J., LEDINGHAM, P. and MERRIFIELD, C. M. (1983) In situ stress measurement in the Carnmenellis granite – II. Hydrofracture tests at Rosemanowes Quarry to depths of 2000m. *Int. J. Rock Mech. Min. Sci.*, 1983, 63-72
2. GEOTHERMAL ENERGY PROJECT REPORT (1987) The creation of hot dry rock systems by combined explosive and hydraulic fracturing. Camborne School of Mines

6.2 Radioactive waste disposal

Isolating radioactive waste is entirely different from other forms of civil engineering, because of the need to prevent radionuclide migration into the environment over very large timescales. Success depends on careful selection of sites and thorough ground investigation. The specific requirements are that there should be:

- minimal groundwater flow through the geological formations surrounding the repository;
- a very long pathway, if any, between the repository and any aquifers or other potential sources of drinking water;
- no likelihood of human intrusion into the repository in search of mineral resources;
- no likelihood of natural disruptive events such as earthquakes which could compromise the containment of the repository.

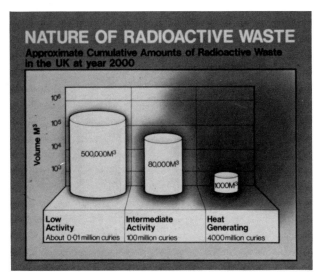

Breakdown of UK waste

It is estimated that about 3.5 million tonnes of radioactive waste will have been produced in the UK by the year 2030. The bulk of this will be low-level waste. If this were all to be disposed of in a single repository it would need a development equivalent to one large coal mine.

A radioactive waste repository could be constructed in a suitable geological environment either deep underground or near the surface, depending on the type of waste. The objective is to isolate the radionuclides while they remain radiologically significant. The most likely mechanism for migration of radionuclides away from a repository is by solution in groundwater, hence the requirement for minimal groundwater movement and a long transit time back to the environment.

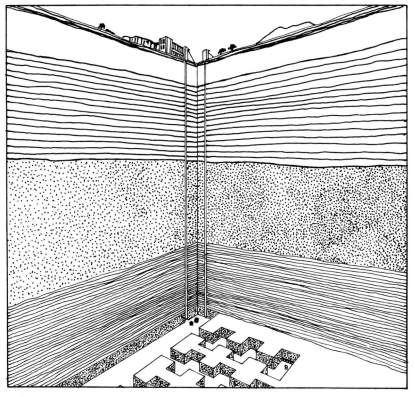

Radioactive waste disposal repository

The engineered component of the repository (known as the near-field) will reinforce the natural properties of the geological environment (the far-field). This provides multi-barrier layers of containment. Concrete surrounding the waste will provide physical and chemical barriers to prevent radionuclides being dissolved and carried away in groundwater. A low-permeability concrete will slow down the movement of water and also make it highly alkaline, reducing the solubility of many radionuclides in the waste. The host rock will slow down migration even further by the process of sorption, when material is removed from water by adhering to the surface of the minerals in the rock.

Again this is a study of interaction between rock mass structure, *in- situ* stress, water flow, heat and time, with all the interactive effects.

Forms of deep repository now being considered include tunnels and caverns, underground or under the sea. Access to the latter could be from the shore or from a facility at sea with a shaft through the seabed.

Waste arriving at a reception area on the surface will be lowered down a shaft or adit to the underground facility where it will be placed in excavated vaults. The vaults will eventually be backfilled with a cement-based grout. This will add stability to the structure, help re-establish groundwater equilibrium, and add a further barrier to radionuclide migration. Finally the whole repository will be sealed.

The degree of long-term radiological protection is calculated by modelling the groundwater flow through rock. This is a complex subject involving, amongst other things, the permeability of the disturbed and undisturbed rock (discussed in Section 2.5), sorption coefficients, solubility coefficients, geochemistry and cement chemistry.

The design requirements of a repository, in addition to the usual civil engineering requirements of underground excavation, also include the radionuclide modelling analysis. The design criteria are very different, although there are some common features, as indicated in the Venn diagram below. Consider how the access requirements or the design life of the repository differ from those of a civil engineering structure. Consider, too, the different contributions made by the structural components of the scheme: the near-field rock around the shaft and excavation, the concrete, the grout, and the far-field rock.

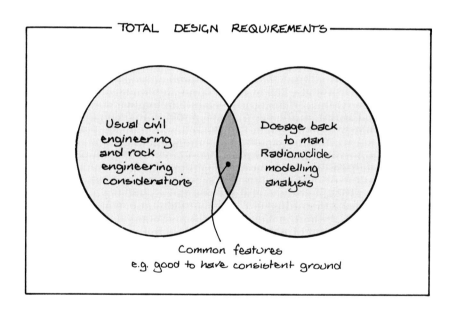

The subject of safe radioactive waste disposal is being researched in many countries, several of which are now constructing underground repositories. Designing a repository is a unique rock engineering exercise, which will use all the design principles described in this book. Each box in the interaction matrix will be relevant. Added to these are the heat which will be emitted in substantial amounts by high-level wastes, the effects of high temperatures on the surrounding rock, and the behaviour of the structure over much longer periods than for conventional development.

Key references

1. RADIOACTIVE WASTE (PROFESSIONAL) DIVISION OF THE DEPARTMENT OF THE ENVIRONMENT (1986) Assessment of best practicable environmental options (BPEOs) for management of low- and intermediate-level solid radioactive wastes. HMSO, London.
2. MILNES, A. G. (1985) Geology and radwaste. Academic Press Geology Series (New York)

6.3　The use of underground space

Civil engineering underground includes a variety of projects, such as road tunnels, large sewer tunnels and clean water schemes (e.g. Dinorwig and Kielder). Similarly, mining engineering activities encompass a wide variety of engineering geometries. In the UK, however, the general motivation for going underground has never been as strong in civil engineering as in mining engineering. There has been relatively little exploitation of underground space, other than for mining or for direct civil engineering infrastructure, because the climate is temperate, much of the surface is relatively flat, and large areas have a deep cover of soil or soft rock. Near-surface hard rock is found mainly in Scotland, Wales and northern England. As more space is needed for development, however, and as energy saving becomes more imperative, the demand for usable space will increase.

In other countries, much more use has been made of underground space. One example is Finland, where the competent granite allows relatively cheap construction for such purposes as housing, swimming pools, water storage, oil storage, frozen-food storage and churches. Similarly, in other parts of the world underground space is much more widely used than in the UK, because it is promoted by the government planning authorities. Another incentive is lack of space, as in Japan, where only 6% of the land area can be used. In both cold and hot countries, there are significant economic advantages in locating housing partially underground.

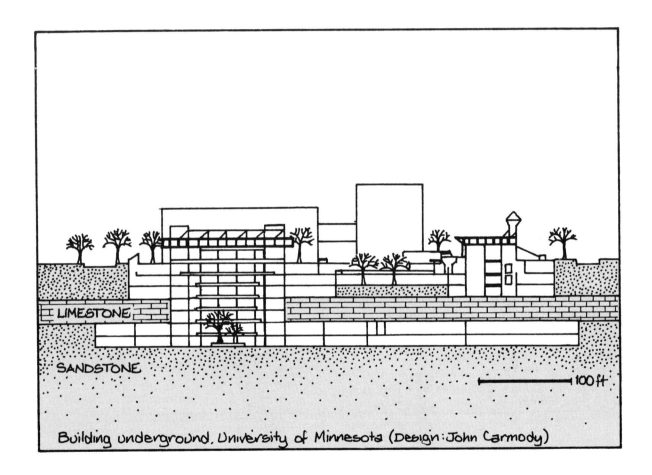

Building underground. University of Minnesota (Design: John Carmody)

The Civil and Mineral Engineering building at the University of Minnesota, illustrated in the section above, demonstrates how a building can be located mainly underground, with the advantage of modifying the effects of seasonal temperature extremes.

In the broad context of civil engineering as a whole, as well as of particular aspects such as inner-city redevelopment, it is important that land-use planning should be fully three-dimensional, to include building depth, as well as height, and the various underground links which allow a building to fulfil its function. The photograph (below) shows the shotcreted and painted walls and roof of the Stockholm Metro. Such transportation systems can be integrated much more directly with housing and shopping, as in France, Japan and Canada, where the Metro exits are directly connected with these facilities. The photograph below shows storage caverns excavated in chalk at Killingholme, UK.

One of the great advantages of underground space is that almost any location and geometry are possible, within the limits of excavation stability.

Stockholm Metro, Sweden

Storage caverns, Killingholme, UK

Because of this greater flexibility and the environmental advantages, applications of underground space are increasing all the time. A key reference, specifically reporting this subject, is the Journal of Tunnelling and Underground Space Technology.

7

Concluding remarks

The photograph of Buddon Wood, UK, on the front cover demonstrates that a rock mass is not likely to be a continuous, homogeneous, isotropic or linearly-elastic material. The principles of rock mechanics which have been outlined in this book, like the principles of soil mechanics or structural mechanics, derive from theories for ideal materials, but their purpose is an understanding of the behaviour of rock in the context of engineering. Rock mechanics is concerned with real (and therefore imperfect), rocks which are additionally disturbed by engineering operations. In other words, rock mechanics provides models of rock mass behaviour; the models can be of varying complexity, of varying types, and to varying scales. They can explain why long-established empirical practices are successful or, indeed, unsuccessful in particular circumstances. Most importantly, these models can form a basis for identifying which engineering techniques may be successfully applied.

Just as each rock mass is unique, so therefore is each rock project. For the project to succeed, skill and judgement based on sound experience are required from the rock engineer. These can only be developed from observing and classifying how rocks behave, and understanding why they do so. Engineering design is concerned with forecasting the behaviour of the ground and the structure built in or on it; each forecast is for an entirely new, if not entirely different, circumstance. The value of rock mechanics lies in providing a framework of understanding for analysis and design. When the type of project is itself new, and there is little or no previous practice, the design has to rely almost entirely on the fundamental principles of rock behaviour.

In this book we have introduced some of the principles of rock mechanics, with the key references providing access to more detailed information. While it is one thing to provide guidance, however, it is another to motivate people to use it. We have attempted here to show the relevance of rock mechanics principles to, and so to encourage their use in, rock engineering practice. Applying these principles in practice is your responsibility. Observe, record, classify and practise – understanding and skill will then develop into the art and science which is rock engineering.

Index

Accuracy 30
anchors
  rock 46
anisotropy 24
aperture
  of discontinuity 9, 15
aquifer 66

Barton's rock mass classification 38
bearing capacity 49
Bienawski's rock mass classification 20, 38
blasting 41, 42, 45, 53
  pre-split 42, 63
  smooth-wall 42
  vibration from 41
block
  caving 48, 60
  failure *see* failure
  size 15, 27, 32, 40
  stability of 44
  structure 46
  theory 44
bolt
  rock 46
brittleness 14
button cutter 43

Cables
  rock 46
cavern 5, 21, 57
Channel Tunnel 47, 48
classification schemes
  rock mass 20, 30, 38, 39, 47
coal
  bump 12
  extraction of 58
cohesion 6, 32, 33
CSIRO gauge 36, 37
compression 9, 14
'connectivity' 23, 64
conservation of vertical load 61
continuum 27, 46

Darcy's Law 22
deformability 6, 7, 9, 20, 24, 26, 33, 49
density (unit mass) 6
dilatometer 20
dimension stone 9
Dinorwig 34, 36, 56, 58
dip 16
dip direction 16, 53
disc cutter 43
discontinuity 7-10, 15-17, 20, 22, 24, 29, 32, 33, 40, 41, 44, 46, 51, 56, 57
  characteristics 15, 31, 32
  frequency 17, 21, 24, 32, 43
  sets of 15, 17, 24, 52
  spacing 15, 21, 32, 56
discontinuum 9, 27
displacement 34, 35, 45, 48
dowels 46
drag picks 43
draw points 60

Elastic constants 7, 50
excavatability 44
excavation
  of rock 40-45, 54, 56
  methods of 45
  surface 53
  underground 54, 57-59, 61
extensometer 34
extraction ratio *R*, 59

Failure
  criteria of 14
  flexural toppling 4, 51
  local shear 49
  of blocks 21, 53–57
  plane 4, 51–53
  stress 44, 54–59
  structural 44, 54
  toppling 4, 51–53
  wedge 4, 17, 51–53
far field 27, 36, 67
filling
  in discontinuity 15
flatjack 36
flow
  of water 11, 23, 26, 28, 57, 64
foundation 4, 49, 50
fracture 6, 7
fracture toughness 43
frequency *see* discontinuity frequency
friction
  angle of 6, 32, 33, 51, 52, 57

Geology
  structural 6, 7
geostatistics 25
geothermal energy 5, 22, 28, 64
Goodman jack 20
ground response curve 35, 45, 47, 48
groundwater flow 22, 66, 67
guniting 46

Hot dry rock *see* geothermal energy
hydraulic
  fracturing 36, 37
  gradient 22
  conductivity
    coefficient of 22

Index tests 7, 8, 29, 30, 38
inhomogeneity 24-26, 29
intact rock 6-9, 14, 20, 22, 24, 33, 40
interaction matrix 28, 44, 45, 63, 64, 67
interactions in rock engineering 10, 11, 28, 63
International Society for Rock Mechanics (ISRM) 2, 15, 29, 36, 59

Kielder project 35, 43, 56, 68
Killingholme 69

Limestone mines
  Black Country 58
longwall 58, 60, 61

Machine
  compression testing 12, 13, 14
micro-seismic events 65
mining 5, 8, 43, 47, 58, 60, 68
Mohr circle 33

Near field 27, 36, 67
New Austrian Tunnelling Method 35, 45

Orientation 15, 44, 51
overbreak 44
overlay 52

Packers 36
peak (breakdown) pressure 36
permeability 7, 9, 15, 22-24, 26, 64
  coefficient of 22
persistence 15, 32, 52
point load
  index value 30
  test 8, 12, 25, 29, 30
Poisson's ratio 6, 7, 50
pole 16
precision 30
pre-split plane 42
pre-splitting 42, 63
pressure
  bulb 49, 50
  hydraulic 45
  peak (breakdown) 36
  residual 36
  water 19
projection (see also stereo-)
  lower-hemisphere 16
  stereographic 16, 17, 51, 52
pumping tests 22

Radioactive waste disposal 5, 22, 64, 66
radionuclide 5, 66, 67
reinforcement 44, 46, 47, 63
Representative Elemental Volume (REV) 26, 27, 29, 48
repository 5, 66, 67
residual pressure 36
resolution 30
roadheaders 47
rock bolts 46
rock bursts 12, 57, 58
rock mass classification 20, 30, 38, 39, 47
Rock Mass Rating (RMR) 20, 38
rock (mass) structure 11, 28, 44, 50, 63, 64
Rock Quality Designation (RQD) 32
rock ripping 9, 40
rock slope stability 17
room and pillar 58, 59
'rose' diagrams 23
roughness 15

Safety
  factor of 59
scale effects 21
scanline 32

Schmidt rebound hammer 25, 30
seepage 15
semivariogram 25
settlement 48
sets *see* discontinuity
shaft 4, 57
shear 9, 14
shearbox
    field 33
shotcrete 46
site investigation 29, 30, 48
slip 32, 33, 46, 52, 55, 57
    rotational 53
slope
    of rock 4, 42
    failure 4, 53
soil mechanics and rock mechanics 10
sonic velocity test 30
sorption 67
stability
    of blocks 44, 57
    of excavations 45, 47
    of slopes 51, 53
stand-up time 47
stereogram 16, 17, 51, 52
stereonet 18, 52
stereoplot 52

stiffness
    compressive 15
    normal 15
    of rock 12
    of testing machine 13
    shear 15
stopes 55
strain
    longitudinal 34
    shear 34
strain gauge 34
strength
    from point load tests 30
    peak 12
    post-peak 12
    tensile 41
    unconfined compressive 6, 8, 10, 12, 14, 25, 41, 43
stress
    circumferential 57
    effective 19
    horizontal 19
    induced 19
    in situ 11, 18, 26, 28, 42, 64
    normal 18, 32, 33, 45, 49, 57
    principal 17, 18, 19, 33, 36, 42, 45, 57, 64
    shear 18, 32, 33, 45
    vertical 18, 37

Stress Concentration Factor (SCF) 59
stress wave 41
strike 16
support 40, 44–48, 56

Tension 9, 14
thermal conductivity 64
time
    effects of 11, 58
trace length 32
tunnel 4, 57
    boring machine (TBM) 9, 43
    linings 47
    portal 62

Underground space 5, 68
USBR gauge 37

Venn diagram 67

Wall strength 15
water jets 43
weathering 58

Young's Modulus 6, 7, 14, 15, 20, 50